p43 Egyptian goal:

p49 annual value method ~~formula~~ ?

⟶ Am Soc of Logistic Engineers ✓

Trip to visit Japan's TPM

p48 Blanchard work

INTRODUCTION TO
TPM

INTRODUCTION TO TPM

TPM

Total Productive Maintenance

Seiichi Nakajima

Publisher's Foreword by Norman Bodek

Originally published by the Japan Institute
for Plant Maintenance

Productivity Press
Portland, Oregon

Originally published as *TPM Nyumon* by the Japan Institute for Plant Maintenance, Tokyo. Copyright © 1984 by Seiichi Nakajima.

English translation copyright © 1988 Productivity Press, Inc.

Productivity Press, Inc.
P.O. Box 13390
Portland, OR   97213-0390
Telephone: (503) 235-0600
Telefax: (503) 235-0909
E-mail: service@ppress.com

Library of Congress Catalog Card Number: 88-61394
ISBN: 0-915299-23-2

Cover design by Gail Graves

Library of Congress Cataloging-in-Publication Data

Nakajima, Seiichi, 1928 –
  Introduction to TPM.

  Translation of: TPM nyūmon.
  Includes index.
  1. Plant maintenance — Management.   I. Title
TS192.N3513   1988   658.2'7   88-61394
ISBN 0-915299-23-2

00 99 98 97   10 9 8 7 6 5 4

Table of Contents

List of Figures

List of Tables

Publisher's Foreword

I am very grateful for the opportunity to publish the English version of *TPM: Introduction to Total Productive Maintenance* by Seiichi Nakajima, Vice Chairman of the Japan Institute of Plant Maintenance. TPM is an innovative approach to maintenance that optimizes equipment effectiveness, eliminates breakdowns, and promotes autonomous operator maintenance through day-to-day activities involving the total workforce. This book introduces those concepts for managers and outlines a three-year program for systematic TPM development and implementation.

The Japanese have a knack for turning good *ideas* into enormously successful *practices*. For example, many Japanese companies applied the ideas of Deming, Juran, and Crosby in innovative day-to-day management and improvement activities to achieve model quality control systems and unparalleled quality records. Similarly, Seiichi Nakajima introduced American maintenance practices in Japan; then, in the early 1970s, he combined those ideas with the concepts of total quality control and total employee involvement to develop Total Productive Maintenance, a system that is revolutionizing plant maintenance around the world.

The changes Nakajima proposes for manufacturing environments are long overdue. TPM promotes group activities throughout the organization for greater equipment effectiveness and trains operators to share responsibility for routine inspection, cleaning, maintenance, and minor repairs with maintenance personnel. Over time, this cooperative effort dramatically increases productivity and quality, optimizes equipment life cycle cost, and broadens the base of every employee's knowledge and skills.

Who can argue with these objectives? After all, the idea behind TPM is not revolutionary — it's just cooperating to get an important job done; for too long, however, we've tolerated some major obstacles to this goal. In many plants operators don't know how to maintain or repair their own equipment, and those who do know are not permitted to do the work because it's someone else's job.

We isolate workers and limit their development by creating exclusive job classifications. Moreover, we thoughtlessly accept the productivity losses that occur when skilled workers are unavailable to repair malfunctioning equipment or treat the first symptoms of imminent failure.

In today's competitive environment we cannot settle for less ambitious goals than the total elimination of breakdowns and other losses and continuous productive maintenance. This means more than periodic overhauls to prevent machine failure. Productive maintenance creatively combines preventive, predictive and maintainability improvement techniques with principles of design-to-life-cycle-cost (DTLCC) to assure reliability in function and ease of maintenance. This is especially important in our increasingly automated manufacturing environments. But this job is too big for a single group of workers and engineers, no matter how skilled they may be; and in attempting it, we lose the productive synergy that develops among people working toward a common goal.

Certainly, every team member has specialized skills, but in a truly cooperative venture, those skills are shared, and everyone grows in understanding and expertise. This approach, practiced by all departments at every stage in the TPM development program, is Japan's unique contribution to the field of plant maintenance.

TPM: Introduction to Total Productive Maintenance is the first book in English on this important subject; a more comprehensive volume on the practical application of total productive maintenance, *TPM Development Program*, will be available from Productivity Press early in 1989. We are proud to make these books available.

I am grateful to Seiichi Nakajima and his staff for their assistance in preparing the manuscripts. I would also like to thank the translator, Keiko Nakamura, and editors Connie Dyer, Camilla England, and Diane Asay for enhancing the original text and making it more accessible for an English-speaking audience. Thanks also to Esmé McTighe, our production manager, and the excellent production team at Rudra Press.

Norman Bodek
Publisher

Preface

I began studying American preventive maintenance (PM) in 1950 and first visited the United States in 1962. Every year since then I have visited American and European manufacturers to study their manufacturing facilities and learn more about their PM systems. Based on my observations, I developed total productive maintenance (TPM) and introduced it in Japan in 1971.

In September 1987 I led a Japanese maintenance study mission to the United States. During the two-week stay, our team made presentations on TPM at several companies and at the Fourth International Maintenance Conference in Cincinnati, sponsored by the Institute of Industrial Engineers. After we completed the study mission, I left the group to speak in Pittsburgh at a TPM Executive Conference organized by Edward H. Hartmann, president of H.B. Maynard Management Institute. More than 150 executives from 80 companies attended those two conferences. I was impressed by the obvious enthusiasm for TPM in the United States and by the number of people eager to learn how to implement it in their companies.

Although planning for the English version of *Introduction to TPM* began two years ago, publication was delayed for several reasons. I believe, however, since American interest in

TPM expanded greatly in the recent past, that now may be the best time for the book to come out.

The basic concepts of TPM and of the TPM development program have not changed in the five years since the book was published in Japan, but the number and variety of industries implementing TPM has increased and the system is being applied outside Japan as well.

For example, the PM Prize for successful implementation of TPM was awarded to 51 factories in the 12 years between 1971 and 1982, and to 65 factories since then. This averages out to slightly more than three companies per year during the first 12 years, but more than three times as many in the last five years. Furthermore, the level of results achieved by companies has improved each year. Worker productivity increased by 60 percent, and accidental breakdowns were reduced to $\frac{1}{100}$ or $\frac{1}{500}$ — close to zero. Now *every* winner is clean enough to be called a "parlor" factory, like Aishin Seiki, the first outstanding winner in this category.

Sixty percent of the 116 plants awarded the prize in the last 17 years are Toyota group companies and their parts suppliers. This demonstrates the close relationship between just-in-time production and TPM. A survey of the past five years' prize-winning factories reveals, however, that while continuing to grow in the automobile industry, TPM is also spreading to other industries, such as the semiconductor, food, pharmaceutical, paper, printing, cement, ceramics, petrochemical, and oil refinery industries, with particularly significant development in the process industries.

Along with this expansion across industries, the scope of TPM implementation *within* companies has grown. It now involves everyone in the R & D and business divisions as well as in manufacturing. In effect, TPM improves the efficiency of the total organization, not just its manufacturing facilities. As the use of automation and unmanned operation increases, it becomes ever more important that facilities function with optimum efficiency. Clearly, to achieve truly efficient production, product development must participate in the TPM effort and the business division must support it logistically with an efficient flow of information and resources.

The movement toward TPM development has also been evident in other countries. Southeast Asia was the first to import it because of geographic proximity. I have conducted seminars in Thailand, China, Korea, and Taiwan, and TPM is being implemented in factories in those countries. (This book was published in Korean in 1985 and in Chinese in 1986.) Communication with European manufacturers has continued since 1970, when the European Federation of National Maintenance Societies (EFNMS) was established. The Japan Institute of Plant Maintenance (JIPM) makes a presentation at every biannual EFNMS Maintenance Conference.

Of the 16 EFNMS member countries, France has made the most progress; currently several French companies are beginning TPM development programs. (This book was published in French in 1987.) Finally, in response to high demand, I have conducted TPM seminars in Brazil for the past two years, and this book will soon be available there. In Japan, we entertain many TPM study teams from these and other countries, which encourages us to believe that TPM is quickly becoming an internationally-recognized system.

Why is this happening? Let me give you my observations on why TPM is applicable to industries in the U.S. and other countries: The most important features of TPM are 1) activities to maximize equipment effectiveness; 2) autonomous maintenance by operators, and 3) company-led small group activities.

Maximizing equipment effectiveness requires the complete elimination of failures, defects and other negative phenomena — in other words, the wastes and losses incurred in equipment operation. This goal is consistent with Philip Crosby's philosophy of zero defects (ZD), an approach to quality management that is gaining popularity throughout the world. Enterprises adopting the ZD philosophy should readily accept TPM.

The second concept, autonomous maintenance by operators, may be less acceptable in some companies, depending upon the prevailing labor organization. Even in Japan there was resistance to this aspect of TPM in plants where operation and maintenance were clearly separated. Although there may be some difficulty introducing TPM in American industries with strong labor unions, the traditional division of labor may change naturally as more companies implement factory automation systems. If labor unions accept these inevitable structural changes, TPM should not be difficult to introduce.

Company-led small group activity, now widely accepted in Japan, is consistent with Likert's participative management model, with Ouichi's Theory Z management, and with Peters' and Waterman's definition of the excellent company in *In Search of Excellence* (New York: Harper & Row, 1982). If such people-oriented management models are taking root in the United States, then American industries may well be ready to implement TPM.

I hope this book will serve as an introductory guide to those planning to implement TPM and that it will help readers find a new way to survive in these times of intense international economic competition.

My thanks to Steven C. Ott, general manager of Productivity Press, and to all the other staff members who helped produce the English version of this book.

Seiichi Nakajima
Vice Chairman
Japan Institute of Plant Maintenance

INTRODUCTION TO
TPM

1

TPM Is Profitable

Total productive maintenance (TPM) is productive maintenance carried out by all employees through small group activities. Like TQC, which is companywide total quality control, TPM is equipment maintenance performed on a companywide basis.

THE NEW DIRECTION IN PRODUCTION

TPM is the new direction in production. In this age, when robots produce robots and 24-hour automated production is a reality, the unmanned factory has become a realistic possibility. In discussing quality control, people often say that quality depends on process. Now, with increasing robotization and automation, it might be more appropriate to say that quality depends on equipment. Productivity, cost, inventory, safety and health, and production output — as well as quality — all depend on equipment.

Production equipment has become unimaginably sophisticated. We see equipment for automation, such as robots and

1

unmanned production; we also see equipment for super-precise processing of micron-size objects and processing that requires speeds, pressures, and temperatures challenging current technology.

Increased automation and unmanned production will not do away with the need for human labor — only operations have been automated; maintenance still depends heavily on human input. Automated and technologically advanced equipment, however, requires skills beyond the competence of the average maintenance supervisor or worker, and to use it effectively requires an appropriate maintenance organization. TPM, which organizes all employees from top management to production line workers, is a companywide equipment maintenance system that can support sophisticated production facilities.

The dual goal of TPM is zero breakdowns and zero defects. When breakdowns and defects are eliminated, equipment operation rates improve, costs are reduced, inventory can be minimized, and as a consequence, labor productivity increases. As Table 1 illustrates, one firm reduced the number of breakdowns to ¹⁄₅₀ of the original number. Some companies show 17–26 percent increases in equipment operation rates while others show a 90 percent reduction in process defects. Labor productivity generally increased by 40–50 percent.

Of course, such results cannot be achieved overnight. Typically, it takes an average of three years from the introduction of TPM to achieve prize-winning results. Furthermore, in the early stages of TPM, the company must bear the additional expense of restoring equipment to its proper condition and educating personnel about the equipment. The actual cost depends on the quality of the equipment and the quality of maintenance. As productivity increases, however, these costs are quickly replaced by profits. For this reason TPM is often referred to as "profitable PM."

Category	Examples of TPM Effectiveness
P **(Productivity)**	• Labor productivity increased: 140% (Company M) 150% (Company F) • Value added per person increased: 147% (Company A) 117% increase (Company AS) • Rate of operation increased: 17% (68% → 85%) (Company T) • Breakdowns reduced: 98% (1,000 → 20 cases/mo.) (Company TK)
Q **(Quality)**	• Defects in process reduced: 90% (1.0% → 0.1%) (Company MS) • Defects reduced: 70% (0.23% → 0.08%) (Company T) • Claims from clients reduced: 50% (Company MS) 50% (Company F) 25% (Company NZ)
C **(Cost)**	• Reduction in manpower: 30% (Company TS) 30% (Company C) • Reduction in maintenance costs: 15% (Company TK) 30% (Company F) 30% (Company NZ) • Energy conserved: 30% (Company C)
D **(Delivery)**	• Stock reduced (by days): 50% (11 days → 5 days) (Company T) • Inventory turnover increased: 200% (3 → 6 times/mo.) (Company C)
S **(Safety/** **Environment)**	• Zero accidents (Company M) • Zero pollution (every company)
M **(Morale)**	• Increase in improvement ideas submitted: 230% increase (36.8 → 83.6/person per year) (Company N) • Small group meetings increased: 200% (2 → 4 meetings/mo.) (Company C)

Table 1. Examples of TPM Effectiveness (Recipients of the PM Prize)

ACHIEVING ZERO BREAKDOWNS:
"THE PARLOR FACTORY"

At the Nishio pump factory of Aishin Seiki, which is called a "parlor factory," you must take off your shoes at the entrance. [In Japan, you never wear your shoes beyond the entrance to a home.] Inside, every section is so clean you may find it hard to believe that you are in a factory that processes metal cuttings. Thus, the nickname "parlor" is quite appropriate. The *zashiki* or "parlor" is the room where the Japanese entertain their guests, and it is usually kept immaculately clean. Although a closer look at the factory reveals cut metal processing for car pump parts, as well as oil and metal filing dust, there are neither oil splatters nor dust on the floor. The floor actually sparkles.

Since 1972, Aishin Seiki has received a prestigious award every five years: the Deming Prize in 1972, the Japan Quality Control Prize in 1977, and the PM Prize in 1982. Aishin Seiki introduced TPM in 1979 with the idea that "no-man" manufacturing (automation) must begin with a "no-dust" workplace. The company's plan to reorganize the factory to improve operation management systems for each product created an opportunity to use the Nishio pump factory as a model "parlor factory."

In addition, a successful worker participation campaign under the slogan, "Let's create our own workplace with our own hands" initially contributed to a cleaner workplace. Workers sacrificed weekends and holidays to help in this effort. At Aishin Seiki, the standards for a clean workplace are based on the 6 S's , the traditional 5 S's (*seiri* — organization, *seiton* — tidiness, *seiso* — purity, *seiketsu* — cleanliness, and *shitsuke* — discipline), plus a sixth S, *shikkari-yarou*, or "let's try hard!" — which means that every person should show initiative and make a special effort.

Workers' efforts at Aishin Seiki eventually bore fruit. Since May 1982 there have been no equipment breakdowns; prior to TPM implementation they numbered more than 700 per month! Furthermore, the current level of quality is extraordinary — a mere eleven defects for every one million pumps produced! Indeed, the plant is now prepared for the unmanned production of the future.

2

TPM
Challenging Limits

Following World War II, the Japanese industrial sectors borrowed and modified management and manufacturing skills and techniques from the United States. Subsequently, products manufactured in Japan became known for their superior quality and were then exported to the Western industrial nations in large quantities, focusing world attention on Japanese-style management techniques.

FROM PM TO TPM

The same has happened in the field of equipment maintenance. More than thirty years have passed since Japan imported preventive maintenance (PM) from the United States. Later adoptions include productive maintenance (PM), maintenance prevention (MP), and reliability engineering. What we now refer to as TPM is, in fact, American-style productive maintenance, modified and enhanced to fit the Japanese industrial environment.

TPM is now well-accepted by the Japanese industrial sector, and is attracting the attention of Western industrial nations, China, and various southeast Asian countries.

FOUR DEVELOPMENTAL STAGES OF TPM

Preventive maintenance was introduced in the 1950's, with productive maintenance becoming well-established during the 1960's. (See Table 2.) The development of TPM began in the 1970's. The period prior to 1950 can be referred to as the "breakdown maintenance" period.

As illustrated in Table 3, the growth of PM in Japan can be divided into the following four developmental stages:

Stage 1: Breakdown Maintenance
Stage 2: Preventive Maintenance
Stage 3: Productive Maintenance
Stage 4: TPM

More recently, both predictive maintenance and equipment diagnostic techniques have attracted considerable attention. These techniques indicate the direction of future PM development.

In a company, TPM is achieved in stages corresponding to the stages of TPM development in Japan between 1950 and 1980. The information in Table 3 is based on data collected in 1976 and 1979 from 124 factories belonging to the JIPM. In three years, the number of factories actively practicing TPM more than doubled. Now, more than one fifth of these factories practice TPM.

Until the 1970's, Japan's PM consisted mainly of preventive maintenance, or time-based maintenance featuring periodic servicing and overhaul. During the 1980's preventive maintenance is rapidly being replaced by predictive maintenance, or condition-based maintenance. Predictive maintenance uses

	1950s	1960s	1970s
ERA	Preventive Maintenance — establishing maintenance functions	Productive Maintenance — recognizing importance of reliability, maintenance, and economic efficiency in plant design	Total Productive Maintenance — achieving PM efficiency through a comprehensive system based on respect for individuals and total employee participation
THEORIES	• PM (Preventive Maintenance) 1951- • PM (Productive Maintenance) 1954- • MI (Maintainability Improvement) 1957-	• Maintenance prevention 1960- • Reliability engineering 1962- • Maintainability engineering 1962- • Engineering economy	• Behavioral sciences • MIC, PAC, and F plans[1] • Systems engineering • Ecology • Terotechnology • Logistics
MAJOR EVENTS	**1951** Toa Nenryo Kōgyō is the first Japanese company to use American-style PM **1953** 20 companies form a PM research group (later the Japan Institute of Plant Maintenance (JIPM)) **1958** George Smith (U.S.) comes to Japan to promote PM	**1960** First maintenance convention **1962** Japan Productivity Association sends mission to U.S. to study equipment maintenance **1963** Japan attends international convention on equipment maintenance (London) **1964** First PM prize awarded in Japan **1965** Japan attends international convention on equipment maintenance (New York) **1969** Japan Institute of Plant Engineers (JIPE) established	**1970** International convention on equipment maintenance held in Tokyo (co-sponsored by JIPE and JMA) **1970** Japan attends international convention on equipment maintenance sponsored by UNIDO[2] (West Germany) **1971** Japan attends international convention on equipment maintenance (Los Angeles) **1973** UNIDO sponsors maintenance repair symposium in Japan **1973** Japan attends international terotechnology convention (Bristol, England) **1974** Japan attends EFNMS[3] maintenance congress **1976** Japan attends EFNMS maintenance congress **1978** Japan attends EFNMS maintenance congress **1980** Japan attends EFNMS maintenance congress

[1] Management for Innovation and Creation (MIC); Performance Analysis and Control (PAC); Foreman Plan (F Plan)
[2] United Nations Industrial Development Organization (UNIDO)
[3] European Federation of National Maintenance Societies (EFNMS)

Table 2. Development of PM in Japan

		1976	1979
Stage 1	Breakdown maintenance	12.7%	6.7%
Stage 2	Preventive maintenance	37.3%	28.8%
Stage 3	Productive maintenance	39.4%	41.7%
Stage 4	TPM	10.6%	22.8%

Table 3. The Four Developmental Stages of PM and the Current Situation in Japan

modern monitoring and analyzing techniques to diagnose the condition of equipment during operation — to identify the signs of deterioration or imminent failure.

DEFINITION AND DISTINCTIVE FEATURES OF TPM

TPM is often defined as "productive maintenance involving total participation." Frequently, management misconstrues this to mean workers only and assumes that PM activities are to be carried out autonomously on the floor. To be effective, however, TPM must be implemented on a companywide basis. Unfortunately, some firms abandon TPM because they fail to support workers fully or involve management.

A complete definition of TPM includes the following five elements:

1. TPM aims to maximize equipment effectiveness (overall effectiveness).
2. TPM establishes a thorough system of PM for the equipment's entire life span.
3. TPM is implemented by various departments (engineering, operations, maintenance).
4. TPM involves every single employee, from top management to workers on the floor.

5. TPM is based on the promotion of PM through *motivation management*: autonomous small group activities.

The word "total" in "total productive maintenance" has three meanings that describe the principal features of TPM:

1. *Total effectiveness* (referred to in point 1 above) indicates TPM's pursuit of economic efficiency or profitability.
2. *Total maintenance system* (point 2) includes maintenance prevention (MP) and maintainability improvement (MI) as well as preventive maintenance.
3. *Total participation of all employees* (points 3, 4, and 5) includes autonomous maintenance by operators through small group activities.

The first principal feature of TPM, "total effectiveness" or "profitable PM," is also emphasized in predictive and productive maintenance. The second feature, a "total maintenance system," is another concept first introduced during the productive maintenance era. It establishes a maintenance plan for the equipment's entire lifespan and includes maintenance prevention (MP: maintenance-free design), which is pursued during the equipment design stages. Once equipment is assembled, a total maintenance system requires preventive maintenance (PM: preventive medicine for equipment) and maintainability improvement (MI: repairing or modifying equipment to prevent breakdowns and facilitate ease of maintenance). The last feature, "autonomous maintenance by operators" (small group activities), is unique to TPM (Figure 1).

In American-style PM, the maintenance department is generally responsible for carrying out PM. This reflects the concept of division of labor, an important feature of American labor unions. Japanese-style PM, or TPM, on the other hand, relies on everyone's participation, particularly autonomous maintenance by operators.

	TPM features	Productive Maintenance features	Preventive Maintenance features
Economic efficiency (profitable PM)	O	O	O
Total system (MP-PM-MI)*	O	O	
Autonomous maintenance by operators (small group activities)	O		

TPM = Productive Maintenance + small-group activities

*MP = maintenance prevention
 PM = preventive maintenance
 MI = maintainability improvement

Figure 1. The Relationship Between TPM, Productive Maintenance, and Preventive Maintenance

If a company is already practicing productive maintenance, TPM can be adopted easily by adding autonomous maintenance by operators to the existing system. If a company has not yet implemented preventive or productive maintenance, however, a sudden shift from breakdown maintenance to TPM will be extremely difficult, although not impossible.

STRIVING FOR OVERALL EQUIPMENT EFFECTIVENESS

The object of production improvement activities is to increase productivity by minimizing input and maximizing output. More than sheer quantity, "output" includes improving quality, reducing costs, and meeting delivery dates while increasing morale and improving safety and health conditions, and the working environment in general.

The relationship between input and output in production activities can be illustrated in a matrix (Figure 2). *Input* consists of labor, machine, and materials, while *output* is comprised of

P Q C D S M

production (P), quality (Q), cost (C), delivery (D), safety, health and environment (S), and morale (M).

Correlating these factors in terms of equipment maintenance demonstrates clearly that all aspects of PQCDSM are related to output. With increasing robotization and automation, the more the production process shifts from workers to machines, the larger the role played by the equipment itself in controlling output, or PQCDSM. Productivity, quality, cost, and delivery, as well as safety and health, environment, and morale all depend on the condition of equipment.

TPM strives to maximize output (PQCDSM) by maintaining ideal operating conditions and running equipment ef-

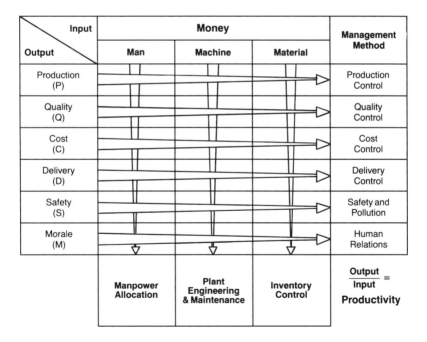

Figure 2. Relationship Between Input and Output in Production Activities

fectively. A piece of equipment that suffers a breakdown, experiences periodic speed losses, or lacks precision and produces defects is not operating effectively.

The life cycle cost (LCC) (the cost incurred during the equipment's lifespan) required to maintain equipment at its optimal level is limited. TPM strives to achieve overall equipment effectiveness by *maximizing* output while *minimizing* input, *i.e.*, LCC.

To achieve overall equipment effectiveness, TPM works to eliminate the "six big losses" that are formidable obstacles to equipment effectiveness. They are:

Down time:

1. Equipment failure — from breakdowns
2. Setup and adjustment — from exchange of die in injection molding machines, etc.

Speed losses:

3. Idling and minor stoppages — due to the abnormal operation of sensors, blockage of work on chutes, etc.
4. Reduced speed — due to discrepancies between designed and actual speed of equipment

Defect:

5. Process defects — due to scraps and quality defects to be repaired
6. Reduced yield — from machine startup to stable production

ZD AND TPM: DEFECT PREVENTION SYSTEMS

The ZD (zero defects) movement came to Japan from the United States in 1965, when ZD campaigns were widespread there. In 1979, Philip Crosby, the creator of ZD, published the best-selling *Quality is Free* (New York: McGraw-Hill, 1979), in which he defined quality as "conformity to requirements" and called for the companywide implementation of ZD or "quality management." His four principles of quality management are listed in Table 4. The goal of zero defects is to create a means of promoting prevention — an essential element in the pursuit of quality.

Zero defects and productive maintenance have a common philosophy. While zero defects strives to prevent defects, productive maintenance in Japan has emphasized the importance of preventing breakdowns for over 30 years. Since equipment failure is a type of defect, both ZD and PM, in effect, are preventive systems aimed at eliminating defects.

THE TOYOTA PRODUCTION SYSTEM AND TPM

The Toyota production system is well-known both in Japan and abroad. It has been argued that Japanese automobile manufacturers achieve their superior productivity through the practice of ZD by auto and parts manufacturers.

Unlike a traditional quality effort based on defect discovery, Japanese-style ZD is based on defect prevention. In Japan, the operators themselves are the inspectors responsible for quality assurance.

1. The definition of quality is conformance to requirements
2. The system of quality is prevention
3. The performance standard is zero defects
4. The measurement of quality is the price of nonconformance

From Philip B. Crosby, *Quality Without Tears* (New York: McGraw-Hill Book Co., 1984)

Table 4. The Four Absolutes of Quality Management

ZD is considered a significant factor in the success of a just-in-time production system as well. Indeed, if there were defects among supplied parts, just-in-time production as well as zero inventory production would be impossible. Thanks to ZD, parts inventories in Japan are usually held from two hours to two days, compared to Western averages of up to 10 days.

Figure 3 illustrates the relationship between TPM and the principal features of the Toyota production system. According to its creator, Ohno Taiichi, the Toyota production system is based on the absolute elimination of waste, namely the elimination of defects and inventories in just-in-time production to produce "the necessary objects, when needed, in the amounts needed." The Toyota production system strives to attain zero defect and zero inventory levels, which, in essence, is ZD.

As illustrated on the right side of the figure, the purpose of TPM is to eliminate the six big losses. This corresponds to the absolute elimination of waste in the Toyota production system.

In striving for zero breakdowns, TPM promotes defect-free production, just-in-time production, and automation. It is safe to say that without TPM, the Toyota production system could not function. The fact that Toyota-related companies are rapidly implementing TPM confirms its importance in the Toyota production system.

RELATIONSHIP BETWEEN TPM, TEROTECHNOLOGY, AND LOGISTICS

Since 1972, the European Federation of National Maintenance Societies (EFNMS) has sponsored a biannual international conference. As one of the member countries, Japan is a regular participant.

During the fifth international EFNMS conference held in Opatija, Yugoslavia, Dennis Parkes (originator of terotechnology in the United Kingdom) gave a keynote address entitled

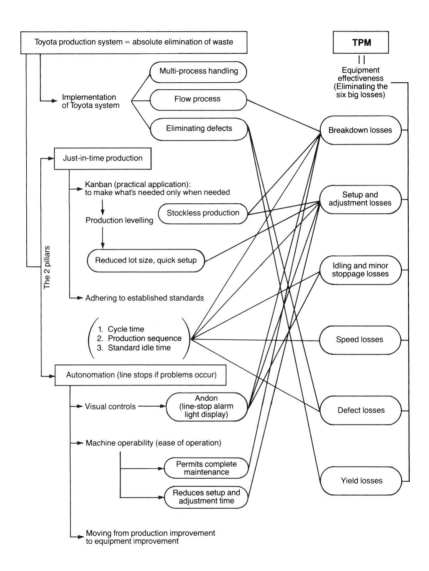

Figure 3. Toyota Production System and TPM

"Progress Report on Terotechnology." In his presentation, Mr. Parkes highlighted the growing importance of terotechnology on an international level and remarked that TPM in Japan and logistics in the United States have the same goals as terotechnology.

"Terotechnology" is a new term coined in the U.K. in 1970. According to the definition endorsed by the British Standards Institution, it is "a combination of management, financial, engineering, and other practices applied to physical assets in pursuit of economic life-cycle costs (LCC). Its practice is concerned with the specification and design for reliability and maintainability of plant machinery, equipment, buildings, and structures, with their installation, commissioning, maintenance, modification, and replacement, and with feedback of information on design, performance, and costs."

TPM in Japan aims to maximize equipment effectiveness. In effect, this is the same as terotechnology's goal of attaining an economic life cycle cost. Actually, the United States Department of Defense first proposed the concept of economic life cycle cost and in 1966 embarked on the development of a life cycle cost program. Moreover, since 1976, the Department of Defense has been basing its procurement contracts for weapons and other large-scale systems on LCC.

"Logistics" is an old military term referring to support for the front line through the procurement, storage, transportation, and maintenance of manufactured goods and systems. Current methods of logistics have updated old notions of the life cycle of goods and equipment through the concepts of LCC, reliability engineering, and maintenance engineering.

Although it is true that TPM, terotechnology, and logistics have economic LCC as a common goal, they differ in terms of the precise target and the location of responsibility. (See Figure 4.) Logistics targets an extremely wide field, including manufactured goods, systems, programs, information, and equipment. Focusing only on equipment (available assets), terotechnology involves the equipment supplier, engineering firms, and the equipment user, while TPM is practiced only by the equipment user.

To promote TPM (the attainment of economic life cycle cost) in Japan, we must broaden its field of application beyond equipment users, as terotechnology and logistics have done.

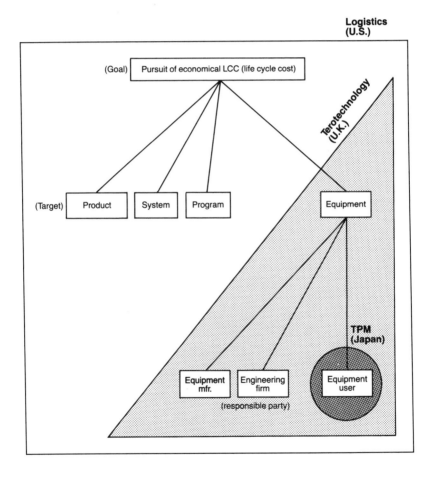

Figure 4. Relationship Between TPM, Terotechnology, and Logistics

3

Maximizing
Equipment Effectiveness

If we are told that equipment effectiveness at Plant X is more than 85 percent, we might reasonably assume that the equipment is being operated efficiently and effectively. But what method of calculation was used to determine the rate of equipment effectiveness and on what data were the calculations based? Many companies use the term "rate of equipment effectiveness," but their methods of calculation vary widely.

EQUIPMENT USED AT HALF ITS EFFECTIVENESS

Often, what is referred to as the rate of equipment effectiveness is actually the operating rate or *availability*.

Measuring Availability

The *operating rate* is based on a ratio of operation time, excluding downtime, to loading time. The mathematical formula for this is:

$$\text{Availability} = \frac{\text{operation time}}{\text{loading time}}$$
$$= \frac{\text{loading time} - \text{downtime}}{\text{loading time}}$$

In this case, *loading time*, or the available time per day (or month), is derived by subtracting the planned downtime from the total available time per day (or month). *Planned downtime* refers to the amount of downtime officially scheduled in the production plan, which includes downtime for scheduled maintenance and management activities (such as morning meetings). For example, assume the working shift per day is eight hours, or 480 minutes. If the planned downtime per day is 20 minutes, then the loading time per day will be 460 minutes.

The *operation time* is derived by subtracting equipment downtime (non-operation time) from loading time; in other words, it refers to the time during which the equipment is actually operating. Equipment downtime involves equipment stoppage losses resulting from failures, setup/adjustment procedures, exchange of dies, etc. For example, again assume a loading time per day of 460 minutes. If downtime per day were composed of breakdowns (20 minutes), setup (20 minutes) and adjustment (20 minutes), or a total of 60 minutes, the operation time per day would be 400 minutes. In this case, the availability or operating rate would be calculated as follows:

$$\text{Availability} = \frac{400 \text{ minutes}}{460 \text{ minutes}} \times 100 = 87\%$$

Accurate Data Is Essential

If the raw data collected in the workplace were accurate, 87 percent availability would be a reliable figure; however, the accuracy of records on actual equipment operation varies depending on the company. Often, such figures are not even recorded. Some managers feel that the time workers spend recording data is wasted and should be used for operational procedures. Minimal operation records must be kept, however, and the recording procedures should be simple and expedient.

Assume, as in the example above, that the planned downtime is 20 minutes, and the recorded downtime 60 minutes, supposedly caused by breakdowns (20 minutes), setup (20 minutes), and adjustments (20 minutes). It is difficult to determine the accuracy of these recorded times. Obviously, there is no need to measure times to the second, but in practice, records often vary from the actual elapsed time by as much as ten minutes. Some companies do not even record equipment failure downtime unless it exceeds thirty minutes. This is an unsound practice. Operation times based on such crude data, in which failure downtime of ten or twenty minutes goes unrecorded, can only lead to crude management as well.

If we want to practice "profitable TPM" and pursue optimal equipment effectiveness, the following two factors are crucial. First, we must keep accurate equipment operation records so that the appropriate management and controls can be provided (with narrower targets); and second, we must devise a precise scale for measuring the equipment operation conditions.

A Broader Range of Factors Must Be Considered

Equipment operation conditions are not reflected accurately when they are based solely on the availability (operation time ratio) figure mentioned above. Of the six big equipment losses, only downtime losses are calculated to determine availability. Other equipment losses such as speed and defect losses are not accounted for. To represent actual equipment operating conditions accurately, all six equipment losses must be included in the calculations.

As shown in Figure 5, TPM includes all six of the big equipment losses in its calculations. It measures overall equipment effectiveness by multiplying availability and performance efficiency by the rate of quality products. This measure of overall equipment effectiveness combines the factors of time, speed, and quality of the equipment operation and measures how these factors can increase added value.

Performance Efficiency

Performance efficiency is the product of the operating speed rate and the net operating rate. The *operating speed rate* of equipment refers to the discrepancy between the ideal speed (based on equipment capacity as designed) and its actual operating speed. The mathematical formula for the operating speed rate is:

$$\text{Operating speed rate} = \frac{\text{theoretical cycle time}}{\text{actual cycle time}}$$

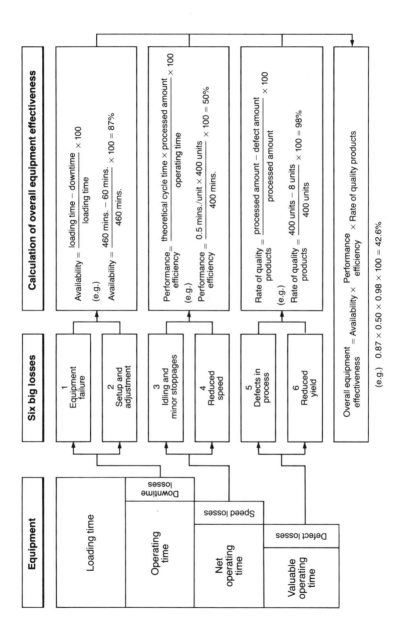

Figure 5. Overall Equipment Effectiveness and Goals

For example, if the theoretical (or standard) cycle time per item is 0.5 minutes and the actual cycle time per item is 0.8 minutes, the calculations would go as follows:

$$\text{Operating speed rate} \ = \ \frac{0.5 \text{ minutes}}{0.8 \text{ minutes}} \ \times \ 100 \ = \ 62.5\%$$

The *net operating rate* measures the maintenance of a given speed over a given period. This figure cannot tell us, however, whether the actual speed is faster or slower than the design standard speed. It measures whether an operation remains stable despite periods during which the equipment is operated at a lower speed. It calculates losses resulting from minor recorded stoppages, as well as those that go unrecorded on the daily logs, such as small problems and adjustment losses:

$$\text{Net operating rate} \ = \ \frac{\text{actual processing time}}{\text{operation time}}$$

$$= \ \frac{\text{processed amount} \ \times \ \text{actual cycle time}}{\text{operation time}}$$

For example, if the number of processed items per day is 400, the actual cycle time per item is 0.8 minutes and the operation time is 400 minutes:

$$\text{Net operation rate} \ = \ \frac{400 \text{ items} \ \times \ 0.8 \text{ minutes}}{400 \text{ minutes}} \ \times \ 100 \ = \ 80\%$$

$100 \ - \ $ net operation rate, which is 20% $=$ losses caused by minor stoppages

Let us calculate the performance efficiency:

Performance efficiency = net operation rate × operating speed rate

$$= \frac{\text{processed amount} \times \text{actual cycle time}}{\text{operation time}} \times \frac{\text{ideal cycle time}}{\text{actual cycle time}}$$

$$= \frac{\text{processed amount} \times \text{ideal cycle time}}{\text{operation time}}$$

$$= \frac{400 \text{ (items)} \times 0.5 \text{ minutes}}{400 \text{ minutes}} \times 100 = 50\%$$

or $(0.625 \times 0.80 \times 100 = 50\%)$

If the rate of quality products is 98 percent, then the overall equipment effectiveness is as follows (see also Table 5):

Overall equipment effectiveness =

Availability × performance efficiency × rate of quality products

or $(0.87 \times 0.50 \times 0.98 \times 100 = 42.6\%)$

Even though the availability is 87 percent, the overall equipment effectiveness, when actually calculated, is not even 50 percent, but an astonishingly low 42.6 percent. The data used in these examples are representative of the average company. In essence, the numbers reveal that equipment was being used at only half its effectiveness.

A: Working hours per day = 60 minutes × 8 hours = 480 minutes
B: Planned downtime per day (downtime accounted for in the production schedule for scheduled maintenance, or for management purposes such as morning meetings) = 20 minutes
C: Loading time per day = A − B = 460 minutes
D: Stoppage losses per day (breakdowns — 20 minutes; setup — 20 minutes; adjustment — 20 minutes) = 60 minutes
E: Operating time per day = C − D = 400 minutes
G: Output per day = 400 items
H: Rate of quality products: 98%
I: Ideal cycle time: 0.5 minutes/item
J: Actual cycle time = 0.8 minutes/item
Therefore,
F: Actual processing time = J × G = 0.8 × 400
T: Availability = E/C × 100 = 400/460 × 100 = 87%
M: Operating speed rate = I/J × 100 = 0.5/0.8 × 100 = 62.5%
N: Net operating rate = F/E × 100 = (0.8 × 400)/400 × 100 = 80%
L: Performance efficiency = M × N × 100 = 0.625 × 0.800 × 100 = 50%
Overall equipment effectiveness = T × L × H × 100 = 0.87 × 0.50 × 0.98 × 100 = 42.6%

Table 5. Overall Equipment Effectiveness Calculations

Based on our experience, the ideal conditions are:

- Availability...greater than 90%
- Performance efficiency...greater than 95%
- Rate of quality products...greater than 99%

Therefore, the ideal overall equipment effectiveness should be:

$$0.90 \times 0.95 \times 0.99 \times 100 = 85 + \%$$

This figure is not just a remote goal. All the PM prize-winning companies have an equipment effectiveness greater than 85 percent.

Table 6 presents overall equipment effectiveness calculations made at a representative plant. Following this example, calculate the overall equipment effectiveness of your own company. Be prepared for a figure that is considerably lower than you expected. The lower the present overall equipment effectiveness, however, the more untapped potential your company possesses.

For example, if the present overall equipment effectiveness at your firm were 50 percent and you used TPM to raise it to 85 percent, the difference would be a 35 percent increase. A ratio of 35 to 50 percent would represent a 70 percent increase in overall equipment effectiveness.

An increase in overall equipment effectiveness produces an increase in productivity. We have presented examples of PM prize-winning firms increasing their productivity by 50 percent, but even a 70 percent increase in productivity is not unattainable.

BREAKDOWNS AND MINOR STOPPAGES IMPEDE AUTOMATION

Murata Kikai is well known as a model factory for the flexible manufacturing system (FMS). Its advanced level of automation is impressive.

Seven machining centers, unmanned carts, automatic pallet changers, automated warehouses, and so on, are controlled by computers. Only two workers were needed during the day to palletize machined parts and transport them to the automated warehouses. The rest of the work is performed 24-hours a day by continuous automated processing.

The work, which is pre-set on pallets, is ordered by the computer to be transported from the automated warehouse by unmanned carts to the automatic pallet changer in the machin-

Process	A Running time (h)	B Planned downtime (h)	C Loading time (h) A − B	D Downtime loss (h)	E Operating time C − D	F Actual processing time (h) J × G	T Operation time ratio E/C × 100(%)	G Output (number of quality products) G_1	G Total (including losses and rework) G	H Quality ratio G_1/G
Internal Finish	17.75	2.3	15.45	—	15.45	10.6	100	4,000	4,018	99.6
	18.58	0.6	17.98	—	17.98	12.8	100	4,800	4,818	99.6
	16.83	0.7	16.13	—	16.13	11.65	100	4,400	4,418	99.6
Total	(59.66)	(3.7)	(55.96)	—	(56.06)	(39.05)	(100)	(14,800)	(14,863)	(99.55)
External Finish	6.5	0.1	6.4	—	6.4	4.62	100	1,187	1,190	99.7
	17.75	0.7	17.5	0.5	16.55	13.8	97.1	3,513	3,542	99.2
	17.75	0.8	16.95	—	16.95	12.86	100	3,297	3,309	99.6
	16.0	1.1	14.9	—	14.9	12.25	100	3,030	3,150	96.2
Total	(58.0)	(1.7)	(56.3)	(0.5)	(54.8)	(43.62)	(99.27)	(11,027)	(11,191)	(98.67)
External Finish	6.5	0.1	6.4	0.2	6.2	3.17	96.9	1,187	1,202	98.8
	17.75	0.7	17.05	—	17.05	9.48	100	3,513	3,593	99.8
	17.75	0.8	16.95	—	16.95	8.95	100	3,297	3,392	97.2
	16.0	1.1	14.9	—	14.9	8.1	100	3,030	3,060	99.0
Total	(58.0)	(1.7)	(56.3)	(0.2)	(55.1)	(29.7)	(99.22)	(11,027)	(11,247)	(98.2)

Process	I Standard cycle time (s) M/C	J Actual cycle time (s) M/C	L Operation performance ratio M × N (%)	M Operation speed ratio 1/J (%)	N Net operation ratio F/E (%)	Number of minor stoppages	Number of exchanges of diamond grindstone	Overall equipment effectiveness T × L × H
Internal Finish	9.5	9.8	66.4	96.9	68.6	2	2	66.1
	9.5	9.8	68.9	96.9	71.2	4	2	68.6
	9.5	9.8	69.9	96.9	72.2	—	2	69.6
Total	(9.5)	(9.78)	(66.9)	(97.07)	(69.7)	(8)	(7)	66.5
External Finish	14	16.6	60.6	84.3	72.2	1	3	60.4
	14	16.65	70.3	84.1	83.4	—	8	67.6
	14	16.45	64.5	85.1	75.9	—	6	64.2
	14	16.15	71.2	86.7	82.2	—	6	68.1
Total	(14)	(16.45)	(66.3)	(85.05)	(78.42)	(1)	(23)	65.6
External Finish	9.5	10.4	46.4	91.3	51.1	—	1	43.6
	9.5	10.5	50.0	90.1	55.6	—	4	49.9
	9.5	10.5	47.5	90.1	52.8	—	3	45.5
	9.5	10.4	49.6	91.3	54.4	—	3	45.1
Total	(9.5)	(10.45)	(48.4)	(90.7)	(53.47)	—	(11)	48.4

Table 6. Calculations Chart for Overall Equipment Effectiveness

ing center. After being processed in the machining center, the work is once again transported to the warehouse by unmanned carts and stored. Currently, the processed cast-iron product requires four manufacturing processes. One lot averages 30-50 items, and approximately 1,300 different types can be produced. For this reason, the factory is a model workplace for the flexible manufacturing system.

This FMS has not yet achieved full automation. Its one shortcoming lies in the area of maintenance. When breakdowns and minor stoppages occur during the night, the entire automated system shuts down. Although operation procedures involving processing, setup, removal, and conveyance have all been successfully automated, maintenance is still difficult to automate.

Murata Kikai's method of increasing labor productivity by performing setup procedures during the day and using automated production at night has been adopted in various industries; however, unless breakdowns and minor stoppages are completely eliminated, further increases in labor productivity will not be realized.

Neither large breakdowns nor minor stoppages can be ignored. Minor stoppages happen for a variety of reasons. Equipment will stop when work clogs at the top of a chute, when a limit switch slips out of position, or when a sensor is alerted because a quality defect has occurred. As in the Aishin Seiki "parlor factory" that eliminated breakdowns, "no-man manufacturing" (automation) starts with "no-dust" workplaces. We often refer to the 5 S's of equipment maintenance, because we can only eliminate equipment failures and minor stoppages and achieve full automation by performing thorough daily maintenance through cleaning, lubricating, bolting, and so on, and through inspections to create a clean, dust-free workplace.

STOP ACCELERATING THE DETERIORATION
OF EQUIPMENT

We like to say that equipment maintenance means maintaining the health of equipment. Preventive medicine has reduced the incidence of disease and increased the human life span significantly. Similarly, preventive maintenance is preventive medicine and health maintenance for equipment.

Figure 6 compares preventive medicine with preventive maintenance. In preventive medicine, emphasis is placed on the prevention of illness, so that disease will not be contracted at all. Proper diet and basic hygiene (*e.g.*, washing hands, gargling) help prevent disease. In addition, periodic health checkups performed by specialists promote early detection and treatment.

Daily equipment maintenance serves the same purpose. By diligently lubricating, cleaning, and performing adjustments (such as bolting) and conducting inspections, deterioration can be prevented and potential equipment failures (disease) averted. Just as people are responsible for their own health, the person using a piece of equipment should be responsible for its health. In other words, daily maintenance is the responsibility of the equipment operator. This is the basic premise behind autonomous maintenance by operators. Furthermore, maintenance personnel, who in effect are "equipment doctors," are responsible for *periodic inspections* (equipment audits functioning as health checkups) and *preventive repairs* (advance replacements functioning as early treatment).

Thus, preventive maintenance decreases the number of breakdowns (equipment disease) and inevitably increases equipment life span.

The practicality of preventive medicine is easy to grasp. The cost of daily prevention and periodic checkups is minimal

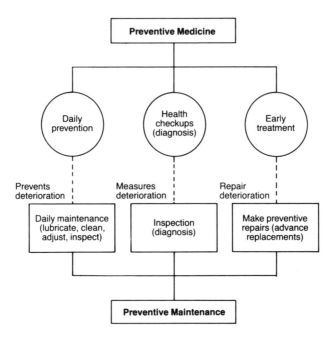

Figure 6. Preventive Medicine for Equipment = Preventive Maintenance

compared to expenses incurred when health care is neglected and when illness leads to hospitalization.

Similarly, it is cheaper to repair the equipment on a preventive basis than to wait until it has completely deteriorated. At that stage the cost of restoring equipment is exorbitant. Oddly enough, however, many companies choose not to practice preventive maintenance or practice it only halfheartedly, even though they understand its importance. Perhaps they are like people who knowingly sacrifice their health and shorten their life spans by overworking and eating and drinking immoderately.

Factories that fail to implement preventive maintenance are, in essence, accelerating the deterioration of their equipment. They are as unlike the "parlor factory" as night is from day. In such factories, powdered dust and chips fly in all directions and lubricants and oil drip while the equipment and floor are littered with dirt, dust, oil, and raw materials.

When dust and dirt adhere to moving parts and sliding surfaces of the machinery, the surfaces are scratched, causing deterioration. And, when lubrication is neglected, excessive friction or burning can result, wasting energy.

A general inspection often reveals that more than half the nuts and bolts are loose. When loosening and deterioration go unattended, they can cause excessive shaking, which encourages abnormal abrasion and triggers further deterioration. Moreover, when plumbing maintenance is inadequate, leaks may develop resulting in excessive waste of precious materials and energy. In factories where such neglect is rampant, sudden failures and minor stoppages are inevitable and common.

Some factory managers have a defeatist attitude: They say, "We can't prevent breakdowns and minor stoppages." Their factories are already in a critical state; breakdowns and minor stoppages have reduced overall equipment effectiveness and reduced productivity. Pressed by the production schedule, these factories do not have the flexibility to implement preventive maintenance. Breakdowns and minor stoppages continue, and conditions go from bad to worse.

At some point, however, unfavorable conditions like these must be stopped. Bad habits and defeatist attitudes can be deeply ingrained in the minds of all the employees — from top management to workers on the floor; they then become part of the company's basic disposition. Middle managers and front line personnel alone cannot change the disposition of the company. Moreover, lukewarm determination will not be enough to change long-standing bad habits. Only when top management is seriously committed to TPM can those habits be discarded and an unfavorable environment altered. Only then is fundamental improvement in a company's disposition possible.

PREVENTIVE MAINTENANCE ALONE CANNOT ELIMINATE BREAKDOWNS

About ten years ago, management at Company N reported that although they had been practicing preventive maintenance for many years, the number of equipment failures had not decreased significantly. They were advised that preventive maintenance alone cannot eliminate breakdowns and encouraged to implement TPM. Taking this action eventually proved successful — they went on to win the PM Prize.

Why can't preventive maintenance by itself eliminate breakdowns?

According to the principles of reliability engineering, the causes of equipment failure change with the passage of time. In Figure 7, breakdowns, or the failure rate, appear on the vertical axis. The failure rate curve is also referred to as the "life span characteristic curve," or the "bathtub" curve (for its characteristic shape). When equipment is new, there is a high failure rate (early failure period), which eventually drops and levels off. Then the failure rate stabilizes at a certain level for a long period of time (accidental failure period). Finally, as equipment approaches the end of its useful life, the failure rate increases once again (wear-out failure period).

At the Hamamatsu factory of the Japan National Railways, Shinkansen (bullet train) cars are serviced. According to the staff there, the failure rate for bullet train cars exactly mirrors the life span characteristic curve. The Tokaido Shinkansen began its service in 1964, the year the Olympics were held in

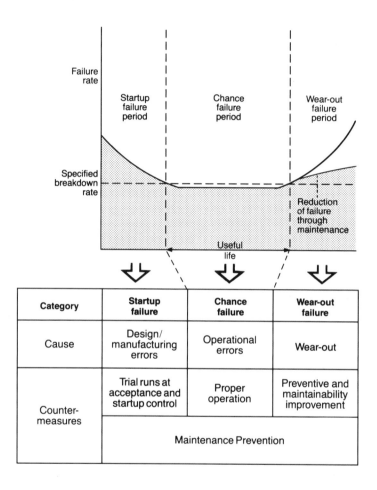

Figure 7. Lifespan Characteristics and Breakdown Countermeasures

Tokyo. At that time, newspapers frequently ran articles such as: "The Shinkansen Fails Again." Breakdowns occurring at that time corresponded to the early period failure.

Early period failures, accidental failures, and wear-out failures each have different causes, as illustrated in Figure 7.

Therefore, to achieve successful results each type of break-down must be treated by different countermeasures.

Causes of early period failure are design and manufacturing errors. To combat them, the design department must conduct test runs at the earliest stage. Furthermore, maintainability improvement should be pursued to discover and treat weaknesses in design and manufacturing.

Accidental failures are caused primarily by operation errors, so the most effective countermeasure is to ensure that operators use equipment properly.

Wear-out failures are due to the limited natural life span of equipment parts. Equipment life can be extended by preventive maintenance and by maintainability improvement (through changes in design). This will reduce the wear-out failure rate.

Maintenance prevention is an effective countermeasure for all three types of breakdowns. A maintenance-free equipment design must be incorporated at the planning/design stage to prevent early period, accidental, and wear-out failures.

When the life cycle of equipment is considered in this manner, it becomes obvious that preventive maintenance alone cannot eliminate breakdowns. The success of TPM depends upon the cooperation of all departments. Maintenance as well as design/planning and operations departments should be involved in the elimination of breakdowns.

FIVE COUNTERMEASURES FOR ZERO BREAKDOWNS

Ideally, breakdowns can be eliminated entirely through maintenance prevention (MP), or the adoption of maintenance-free design. The condition of most factory equipment, however, is far from this ideal.

The first step toward improvement is to eliminate failures in equipment currently being operated. The experiences gained from this endeavor can then be fed back to design better equipment. Gradually the equipment will approach the ideal.

According to the Japanese Industrial Standards (JIS), "a failure results in loss of a standard function in a certain object (*e.g.*, system, machine, part)." This "loss of a standard function" indicates that machine failures are not limited to unexpected breakdowns that lead to a complete stoppage. Even when equipment is running, deterioration can cause various losses, such as the loss of a standard function, longer and more difficult setup/adjustments, frequent idling and minor stoppages, and reduction in processing speed and cycle time. Such losses must be treated as failures.

Unexpected breakdowns with complete stoppage are called "function-loss failures" while those involving equipment deterioration despite continued operation are called "function-reduction failures."

Breakdowns represent the tip of the iceberg. We tend to become overly concerned with breakdowns and serious defects because they are so obvious, and there certainly are cases in which a single defect causes a breakdown. Small defects, however, such as dirt, dust, abrasion, loosening, scratches, and warping — which may seem insignificant on their own — are the real problem.

These small defects can suddenly become large. Sometimes they overlap to create a stronger effect, triggering both function-loss and function-reduction failures. As the saying goes, "even one match can cause a fire," so it is important to stamp out defects while they are small. This is the fundamental concept behind preventive maintenance.

Defects that go undetected and untreated are called "hidden defects." If they remain untreated, they will trigger breakdowns. Therefore, it is important to expose hidden defects and restore optimal conditions.

To eliminate failures we must expose hidden defects and treat equipment before it breaks down. The following five concrete countermeasures help eliminate failures:

1. Maintaining well-regulated basic conditions (cleaning, lubricating, and bolting)
2. Adhering to proper operating procedures
3. Restoring deterioration
4. Improving weaknesses in design
5. Improving operation and maintenance skills

Figure 8 illustrates the relationships between these five countermeasures. Breakdowns often occur because people fail to implement simple measures. As this figure illustrates, breakdowns can be eliminated by carrying out simple procedures in a simple manner.

To ensure that simple procedures are performed thoroughly, both the operations and maintenance departments must understand each other's role and cooperate, like the wheels of a car. They must be willing to adjust their points of view and behavior and fulfill their respective duties. Everyone involved with equipment operation or maintenance must work to eliminate failures. (Figure 9 shows the division of labor between the operations and maintenance departments in pursuing the goal of zero breakdowns.)

UNLICENSED OPERATION OF AUTOMATED EQUIPMENT

Consider two types of companies: The first type of company possesses a high degree of technical ability. Equipment and dies are designed and manufactured within the company and workers are trained to handle their equipment properly. The other type of company subcontracts the production of its equipment and dies and attaches no importance to equipment

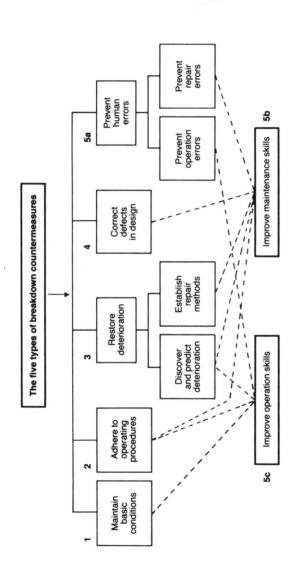

Figure 8. Relationships Between Breakdown Countermeasures

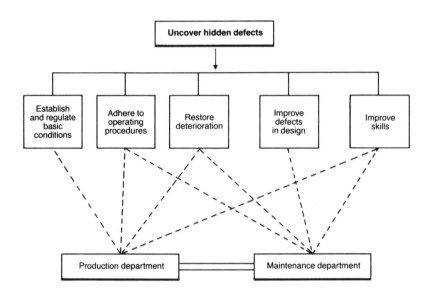

Figure 9. Responsibilities of the Operations and Maintenance Departments

technology. It operates its automated equipment nonchalantly, without technical know-how, like a automobile driver without a license.

Anjo Denki is a subsidiary of Nippondenso with approximately 600 employees. It produces electrical relays specifically for cars. The president of the firm is an expert engineer capable of designing his own equipment. He praises the West German corporate emphasis on technology and manages his company in a similar manner. He is a strong advocate of TPM and the integration of TPM and TQC. In 1978, his company won the PM Prize in Category 2, and in 1981, the Prize in Category 1.

By implementing TPM, Anjo Denki significantly increased its overall equipment effectiveness in such areas as its automated assembly line — even its parent company, Nippondenso, was impressed with the progress. It is company policy to train all of its employees to deal with the equipment. Young production engineers are being trained to develop automatons; operators in the mechanical division, which manufactures dies and special equipment, are studying for the technical qualification exams in engineering to increase their competence; and many regular and part-time employees in the workplace have learned the basics of equipment maintenance and practice autonomous maintenance, taking to heart the principle that all individuals are responsible for their equipment.

From now on, development of efficient equipment will be the key to survival. For example, Yoshida Kōgyō (YKK) develops its own equipment and also dominates the world zipper market.

Like Anjo Denki, most of the companies awarded the PM Prize for TPM implementation possess the technical ability to develop and produce their own equipment and dies. One small company that won the PM Prize in 1981 was particularly impressive. Even though it had fewer than fifty employees, its two veteran engineers developed all the equipment, tools, and dies.

We strongly encourage in-house design and manufacture of equipment and dies, because it helps polish and perfect production techniques and skills. Moreover, keeping up with technological progress is essential to survival.

Companies choosing to ignore TPM are the complete opposite of those described above. They subcontract the design and manufacture of equipment and dies and only a few qualified technicians are on hand to ensure effective equipment operation. But dependence on outside manufacturers for equipment and dies wastes a potential advantage. In addition, a company that has not developed its own equipment, tools, and dies, will also lack the level of technical ability necessary for effective maintenance. Finally, since the life cycle of equipment and dies cannot be controlled within the company, its ability to provide maintenance prevention and corrective maintenance is low.

The rationale behind subcontracting equipment, tools, and dies is to focus energy and resources on production itself rather than on indirectly related processes such as design and development. However, this indifference and sometimes contempt for equipment technology indicates a lack of understanding of its importance in manufacturing today.

Equipment is becoming increasingly sophisticated. Computers as well as hydraulic and electric controls are being used. Companies neglecting technology give their new employees who just graduated from technical high schools only brief training before sending them to the production floor. Obviously, new employees who barely understand the structure and functions of their sophisticated, automated equipment will have difficulty operating it. Defects in process, equipment breakdowns, and accidents are inevitable. Such inadequate use of sophisticated equipment certainly represents a kind of "unlicensed operation" that is a far cry from the level of operations found in companies practicing TPM.

ERRORS IN EQUIPMENT INVESTMENT

Japanese cars are extremely fuel-efficient and their low breakdown rate keeps repair costs down to a minimum. Thus, Japanese cars are more economical than the average American car despite the higher price tag.

In 1983, the author met with a U.S. LCC consultant who, like the author, was a member of the Society of Logistics Engineers (SOLE). Since the life cycle cost (LCC) of Japanese cars is exceptional, the consultant hoped to establish a U.S.-Japan exchange of ideas on this subject.

Certainly since the oil crisis, Japanese car manufacturers have striven to attain high fuel-efficiency and reliability. Furthermore, they have sought to win the international price war by eliminating waste and implementing value engineering (VE). As a result, Japan has surpassed Western nations to achieve the lowest total life cycle costs in terms of purchase price, fuel, and upkeep.

Based on this, the consultant assumed that information on LCC would be widespread in Japanese industry. He was surprised to learn that hardly any Japanese auto manufacturers practice the so-called "Design To LCC" (DTLCC), in which a company commits to pursue economic life cycle cost as early as the design stage. Although the "pursuit of economic LCC" is understood as a concept in Japan and has been put into practice to a limited extent, Japanese manufacturers still lack the "LCC design" procedures that are being developed and actually applied in the U.S.

For other durable consumer goods, such as household appliances and cameras, Japan has the lowest life cycle costs as well. Consumers are the ultimate beneficiaries of a company's pursuit of economic LCC. And, since to increase sales manufacturers are obliged to give consumers what they want, they should pursue economic LCC aggressively. No consumer would object to this.

Why is it, then, that when we talk about the pursuit of economic life cycle in relation to the productive or capital goods of factory production equipment, the discussion becomes vague and ambiguous? With respect to equipment investment, just as with durable consumer goods, an economic LCC will benefit the consumer.

Many companies miscalculate their profits and losses when considering equipment investment. The following example, provided by Professor Senju Shizuo of Keio University, compares the economic effectiveness of two types of paints and illustrates traditional attitudes towards LCC.

To cover a certain area, Paint A would cost $5,000 and last three years. Paint B would cost $15,000 (three times the cost of Paint A) and last six years. Which is more economical?

As illustrated in Figure 10, Paint A would require labor costs of $20,000, resulting in a total figure of $25,000 (including the $5,000 for the paint). Since the life span of the paint is three years, another $25,000 would be needed three years later.

Paint B, on the other hand, would require $15,000 for paint and $20,000 for labor (same as for Paint A). The total figure for Paint B, therefore, would be only $35,000 over a six year period.

Comparison of two types of paints

	Cost of paint	Lifespan	Ratio
Paint A	$5,000	3 years	$1,666/year
Paint B	$15,000	6 years	$2,500/year

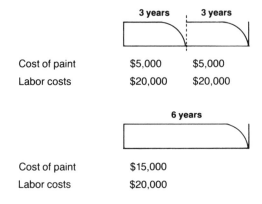

3 years	3 years
Cost of paint	$5,000	$5,000
Labor costs	$20,000	$20,000

| 6 years |
| Cost of paint | $15,000 |
| Labor costs | $20,000 |

(from Senju Shizuo, Session 1, Third Terotechnology Symposium, 1979)

Figure 10. Pursuit of Economic Life Cycle Cost

Comparing the total costs for six years, including labor, Paint B is far more economical — $15,000 less than paint A.

It would be a mistake to purchase paint A based on price alone, saying that paint A costs $5,000 for a life span of three years, or $1,666 per year, while paint B costs $15,000 for a life span of six years, or $2,500 per year. Purchasing the paint is not the only goal of the transaction. If you actually want to have the area painted over a given period of time, clearly, the overall calculations of profits and losses must be based on the total costs for paints A and B, including labor.

The same is true for equipment investment, as the following example will illustrate: Companies A and B have submitted estimates for equipment. As shown in Table 7, company A's estimate is $100,000 and company B's is $70,000. Many companies would compare the two firm's estimates and sign a contract with company B, because "B is cheaper than A."

As the paint example shows, however, it is a mistake to make such a decision based on purchase price alone. It would be more profitable to compare life cycle costs. In other words, we should consider the total costs over a five-year period, including annual maintenance ($30,000 for Company A, $60,000 for Company B) and a 10 percent calculated interest rate.

When comparing long-term economic effectiveness in this manner, we must consider the time value of money. Cost conversions based on factors such as calculated interest must be measured, using tools of engineering economics. There are three methods of comparison available: (1) All monetary earnings and expenses are converted into current values, referred to as the present value method. (2) The final value method is applied at the end of the period of comparison. (3) The annual value method is used to estimate annual expenses.

Here, we shall use the annual and present value methods in comparing the two companies, both of which result in lower figures for Company A, as shown in Table 7. In other words,

	A	B
Purchasing price of equipment	$100,000	$70,000
Annual sustaining cost	$30,000	$60,000
Life	5 years	
Interest rate	10%	
Annual value method	$56,380	$78,466
Present value method	$213,730	$297,460

(Figures for annual and present values based on economic engineering.)

Table 7. A Comparison of Equipment Life Cycle Costs

when the life cycle costs of the two firms are compared over a five-year period, Company A is clearly the better choice.

As mentioned earlier, the U.S. Department of Defense has been approving its contracts for the procurement of large-scale projects based on LCC since 1976. The U.S. Department of Defense used to calculate profitability based simply on a comparison of procurement prices. Over time, however, a rapidly increasing percentage of the defense budget had to be allotted for operation and maintenance costs, rather than for the procurement of new projects. To increase the Department's purchasing power while keeping costs down, procurements were made on an estimated budget based on LCC.

Other examples of the growing use of LCC calculations are businesses that rent or lease construction equipment, computers, and even copying machines. The rental or leasing fees for such equipment are calculated using the annual value method formula, which is based on LCC and includes maintenance costs as well as other service charges.

Through the activities of the American Society for Logistics Engineering, information on life cycle cost is being disseminated internationally — no country can afford to ignore LCC.

THE FIVE TPM DEVELOPMENT ACTIVITIES

The practical details and procedures for using TPM to maximize equipment effectiveness must be tailored to the individual company. Each company must develop its own action plan, because needs and problems vary, depending on the company, type of industry, production methods, and equipment types and conditions.

When members of the Japan Institute of Plant Maintenance consult for a Japanese company, they propose a TPM development schedule designed to meet the company's predisposition, needs, and problems, based on preliminary diagnosis and investigation. There are, however, some basic conditions for the development of TPM that apply in most situations.

Generally, the successful implementation of TPM requires:

1. Elimination of the six big losses to improve equipment effectiveness
2. An autonomous maintenance program
3. A scheduled maintenance program for the maintenance department

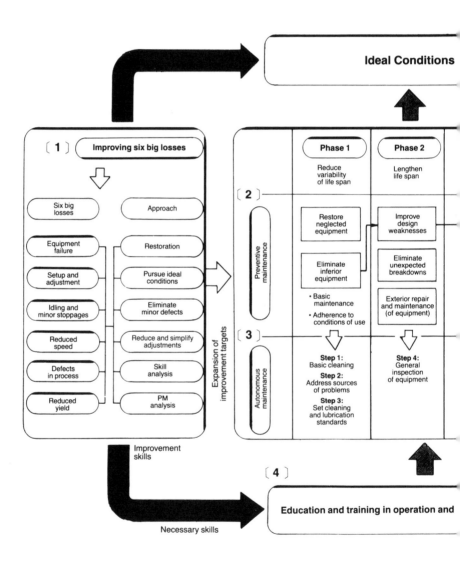

Figure 11. Example of TPM Development

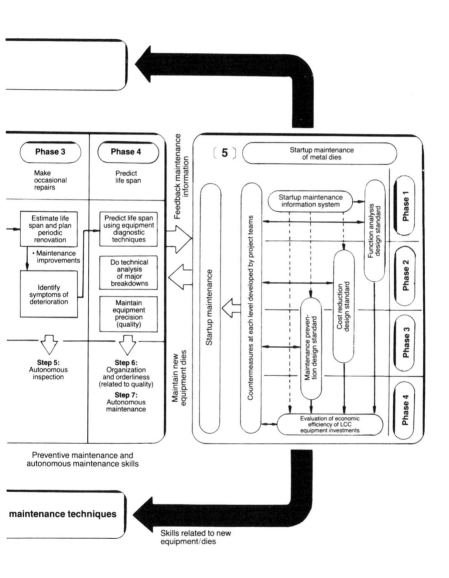

Phase 3

Make
occasional
repairs

Estimate life
span and plan
periodic
renovation

• Maintenance
improvements

Identify
symptoms of
deterioration

Step 5:
Autonomous
inspection

Phase 4

Predict
life span

Predict life span
using equipment
diagnostic
techniques

Do technical
analysis
of major
breakdowns

Maintain
equipment
precision
(quality)

Step 6:
Organization
and orderliness
(related to quality)

Step 7:
Autonomous
maintenance

Feedback maintenance
information

Maintain new
equipment dies

Preventive maintenance and
autonomous maintenance skills

[5] Startup maintenance
of metal dies

Startup maintenance

Countermeasures at each level developed by project teams

Startup maintenance
information system

Maintenance preven-
tion design standard

Cost reduction
design standard

Function analysis
design standard

Evaluation of economic
efficiency of LCC
equipment investments

Phase 1

Phase 2

Phase 3

Phase 4

maintenance techniques

Skills related to new
equipment/dies

   4. Increased skills of operations and maintenance per-
      sonnel
   5. An initial equipment management program

These are the basic TPM development activities. Figure 11
shows how they provided the foundation for a successful TPM
development program.

   These developmental activities constitute the minimal requirements for the development of TPM. They are explained
further in Chapter 4, "The Twelve Steps of the TPM Development Program," and in Chapter 5, "TPM Implementation and
Stabilization." Please refer to *TPM Development Program* (Cam-
bridge: Productivity Press, forthcoming) for a more detailed
explanation.

4

Organizing for
TPM Implementation

Every autumn in Japan, the JIPM sponsors a symposium by representatives of the year's PM prize-winning factories. They explain how their progress and results helped them become prize winners to the top management of other firms currently striving for the PM Prize. Hearing about the experiences of top management in prize-winning companies encourages upcoming managers to pursue this target.

THE TWELVE STEPS OF TPM DEVELOPMENT

It takes at least three years of TPM implementation before prize-winning results can be achieved. Many top managers, enterprising and eager to implement beneficial programs, may say, "If other companies can do it in three years, we can do it in one." Their enthusiasm is commendable, but to invest so little time in an undertaking of this scope can only lead to failure.

Three Requirements for Fundamental Improvement

The goal of TPM is to effect fundamental improvement within a company by improving worker and equipment utilization. To eliminate the six big losses, we must first change people's attitudes and increase their skills. Increasing their motivation (*yaruki*) and competency (*yaruude*) will maximize equipment effectiveness and operation. Such improvements in the quality and functioning of equipment and in one's mental outlook are essential to fundamental corporate improvement.

Yaruba, or work environment, is a third important condition for improvement. We must create a work environment that supports the establishment of a systematic program for implementing TPM. Unless top management takes the lead by seriously tackling this issue, however, the necessary transformation in attitudes, equipment, and the overall corporate constitution will not progress smoothly.

The Three Stages of TPM Development

Table 8 lists the twelve basic steps of a TPM development program. In the *preparation* stage a suitable environment is created by establishing a plan for the introduction of TPM. This preparation stage is analogous to the product design stage, when the details of a product are mapped out and prepared for. It is like the kickoff at the beginning of a ball game — the point at which production begins.

The *implementation* stage is comparable to the production stage for a product. Materials are processed; parts are made, and after inspection, assembled. A final inspection completes the manufacturing process. This period is called the *stabilization* stage.

One of the most effective ways to raise morale and achieve positive results is to give employees a goal to strive for. For

Stage	Step	Details
Preparation	1. Announce top management decision to introduce TPM	Statement at TPM lecture in company; articles in company newspaper
	2. Launch education and campaign to introduce TPM	Managers: seminars/retreats according to level General: slide presentations
	3. Create organizations to promote TPM	Form special committees at every level to promote TPM; establish central headquarters and assign staff
	4. Establish basic TPM policies and goals	Analyze existing conditions; set goals; predict results
	5. Formulate master plan for TPM development	Prepare detailed implementation plans for the five foundational activities
Preliminary implementation	6. Hold TPM kick-off	Invite clients, affiliated and subcontracting companies
TPM implementation	7. Improve effectiveness of each piece of equipment	Select model equipment; form project teams
	8. Develop an autonomous maintenance program	Promote the *Seven Steps;* build diagnosis skills and establish worker certification procedure
	9. Develop a scheduled maintenance program for the maintenance department	Include periodic and predictive maintenance and management of spare parts, tools, blueprints, and schedules
	10. Conduct training to improve operation and maintenance skills	Train leaders together; leaders share information with group members
	11. Develop early equipment management program	MP design (maintenance prevention); commissioning control; LCC analysis
Stabilization	12. Perfect TPM implementation and raise TPM levels	Evaluate for PM prize; set higher goals

Table 8. The Twelve Steps of TPM Development

the Japanese, this goal is winning the PM Prize. Therefore, in preparing for implementation it is important to determine how long it will take to achieve specific TPM goals and to develop a master schedule.

We can estimate roughly how long it will take from initial preparation to the achievement of prize-winning results. Depending on the size of the company, level of technology, management standards and current level of PM, the preparation stage will last three to six months. We allow at least that much time for product design to create better products.

Next, it will take two to three years to complete the implementation process. It is crucial to budget sufficient time for this stage, otherwise the "product" will not be as good as it looks; it will be crudely made, of inferior quality, and short-lived.

During the final period of stabilization, a company must measure actual results achieved against its TPM targets (in Japan, companies are generally evaluated by the PM Prize Committee), and set more challenging goals.

The rest of this chapter describes the five steps of the TPM preparation stage.

STEP 1: ANNOUNCE TOP MANAGEMENT'S DECISION TO INTRODUCE TPM

The first step in TPM development is to make an official announcement of the decision to implement TPM. Top management must inform their employees of this decision and communicate enthusiasm for the project. This can be accomplished through a formal presentation that introduces the concepts, goals, and expected benefits of TPM, and also includes top managers' personal statements to employees on the reasons behind the decision to implement TPM. It may be followed up by printed statements in company bulletins.

It is essential at this point that top management have a strong commitment to TPM and understand what that commitment entails. As mentioned earlier, preparing for implementation means creating a favorable environment for effective change. During this period (as in the product design stage), a firm foundation must be laid so that later modifications (like design changes that can result in delivery delays) will not be necessary.

That is why TPM must be implemented with the unwavering support and firm leadership of top management, even though it depends on *total* employee participation, from top management to front line workers. TPM respects the autonomy of workers, but it promotes autonomous activities only after they have become sufficiently motivated and competent to manage their own activities successfully, and only when a work environment that supports autonomous activities has been created. Establishing that favorable environment is management's primary responsibility at this stage.

During the first two stages of TPM development, management must train employees to deal with the equipment on their own by improving their operation and maintenance skills and promoting autonomous maintenance. This can only happen with a non-authoritarian style of management, but employees must also commit themselves to becoming able to handle and maintain the equipment. Truly independent employees will emerge only when their motivation matches the favorable environment created by management.

It takes considerable time to change people's attitudes and habits, however a 50 percent increase in productivity is certainly within reach. Then, as the shop floor environment improves, employees will find their work more satisfying.

Top management must understand and believe in the concept of TPM before implementing it. Consulting with managers who have successfully implemented TPM or visiting their operations can help eliminate doubt and thus improve the quality of their support for workers on the floor.

STEP 2: LAUNCH EDUCATIONAL CAMPAIGN

The second step in the TPM development program is TPM training and promotion, which should begin as soon as possible after the introduction of the program.

The objective of TPM education is not only to explain TPM, but also to raise morale and soften resistance to change — in this case, to TPM.

Resistance to TPM may take different forms: Some workers may prefer the more conventional division of labor (operators run the equipment, while maintenance workers repair it). Workers on the production line often fear that TPM will increase the work load, while maintenance personnel are skeptical about the operators' ability to practice PM. Moreover, those who are successfully practicing PM may doubt that TPM will provide added benefits.

TPM implementation education should be designed to eliminate resistance and raise morale. In Japan, for example, 2-3 day training retreats by level have been most effective for managers and section chiefs, or for staff engineers and group leaders or foremen. Top management often attend the retreats for upper managers and section chiefs to provide support by their presence.

Floor workers can be trained using slide presentations or other visual materials. This training can be enhanced by inviting supervisors and other managers to TPM small group meetings to relate what they have learned from their own retreats.

During the TPM education stage, a campaign to promote enthusiasm for TPM implementation is usually organized. Japanese companies often use banners, placards, signs, flags, and badges bearing TPM slogans to create a positive environment.

STEP 3: CREATE ORGANIZATIONS TO PROMOTE TPM

Once the introductory education of management-level personnel (section chiefs and above) has been completed, the building of a TPM promotional system can begin.

The TPM promotional structure is based on an organizational matrix, forming horizontal groups such as committees and project teams at each level of the vertical management organization. It is extremely important for the support and successful development of TPM companywide. As illustrated in Figure 12, groups are organized by rank, for example, the TPM promotional committee, departmental and factory promotional committees, and PM circles for the work floor. The integration of top-down, goal-oriented management with bottom-up, small group activities on the work floor is critical.

Traditionally, small group activities, such as QC circles, are organized outside the management structure. When TPM activities cannot be completely integrated within a management structure, these pre-existing QC circles can be used to promote PM activities. Normally, however, autonomous small group activities can be carried on within the existing management structure. New PM circles or groups can be created within this structure by assigning leadership responsibilities to section leaders, group leaders, or foremen on the work floor.

Rensis Likert advocated small group activities as a means of promoting participatory management. Following Likert, JIPM recommends a network of overlapping small groups, organized at every level from top management to the work floor. Each group leader participates as a member in a small group at the next level. In other words, the group leaders

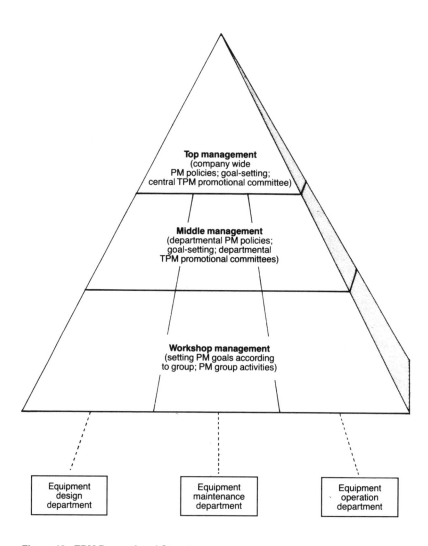

Figure 12. TPM Promotional Structure

serve as a link between levels, facilitating vertical as well as horizontal communication.

Figure 13 is an example of this type of organization for companywide small group activities, illustrating the TPM promotional structure of the 1982 PM Prize winner, Central Motor Wheel Co. At their plants, in addition to small group activities at every level, specialized (professional) committees and project teams have been organized to promote the elimination of the six big losses.

Since TPM is implemented over a three year period, it is important that a TPM promotional headquarters be established and professionally staffed. Serving as general staff to top management, these professionals play a crucial role in the TPM promotional structure. Ideally, these individuals should be full-time employees of the highest caliber, trained in equipment management.

STEP 4: ESTABLISH BASIC TPM POLICIES AND GOALS

The TPM promotional headquarters staff should begin by establishing basic policies and goals. Since it takes at least three years to move toward eliminating defects and breakdowns through TPM, one basic management policy should be to commit to TPM and incorporate concrete TPM development procedures into the medium to long-range management plan.

Although company mottos and slogans are often simply displayed on the walls, concrete basic policies and annual goals of management must be adhered to. Although *policies* may consist of abstract written or verbal statements, the *goals* should be quantifiable and precise, specifying the target (*what*), quantity (*how much*), and time frame (*when*). For example, a basic management policy might be: "To reduce losses by eliminating breakdowns, defects, and accidents while en-

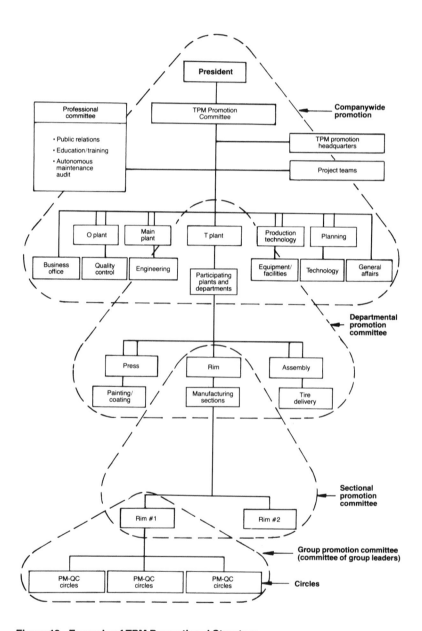

Figure 13. Example of TPM Promotional Structure

hancing the profitability of the company and creating a favorable working environment for all employees." In this statement, the goals of management are clear and succinct, and the basic policy can be expressed in concrete figures, as quantitative goals.

Of course, to totally eliminate breakdowns and defects may be an unattainable goal. Therefore, management should set intermediate goals: for a three-year plan, for example.

To set an attainable goal, the actual level and characteristics of current breakdowns and rate of process defects per piece of equipment must be measured and understood. In some companies this information is not available and we must start by identifying current conditions.

Assume investigations of current equipment reveal 40 breakdowns per month, and a three percent rate of process defects. Using these figures as a bench mark, it is possible to reduce the incidence rate to one-tenth in three years, that is, to four breakdowns a month and a 0.3 percent rate of process defects. To decide what the target levels should be, we must consider both internal and external needs. When this has been established, the three-year goals must be compared to current conditions. Then improvements must be predicted, contributions to the firm's business estimated, and the rate of return on estimated costs of improvements calculated.

Figure 14 is an example of basic TPM policies and goals taken from Tokai Rubber Industries, Ltd. This firm received the PM Prize in 1981 for successfully accomplishing its goals.

Once medium- to long-range goals have been set for the company and the factory, they must be developed further in each department and at each level. Annual goals are determined by managers and supervisors who must ensure that the improvement themes and goals set independently by the floor workers' small groups are consistent with the overall goals of the company.

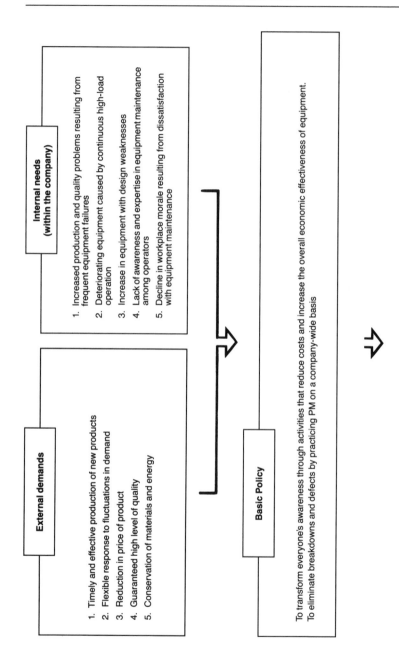

External demands

1. Timely and effective production of new products
2. Flexible response to fluctuations in demand
3. Reduction in price of product
4. Guaranteed high level of quality
5. Conservation of materials and energy

Internal needs (within the company)

1. Increased production and quality problems resulting from frequent equipment failures
2. Deteriorating equipment caused by continuous high-load operation
3. Increase in equipment with design weaknesses
4. Lack of awareness and expertise in equipment maintenance among operators
5. Decline in workplace morale resulting from dissatisfaction with equipment maintenance

Basic Policy

To transform everyone's awareness through activities that reduce costs and increase the overall economic effectiveness of equipment. To eliminate breakdowns and defects by practicing PM on a company-wide basis

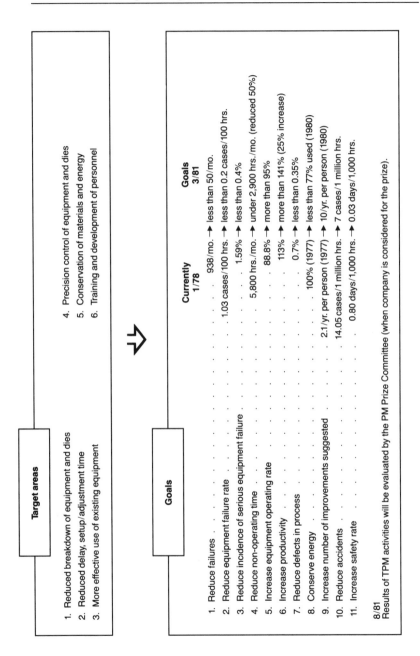

Target areas

1. Reduced breakdown of equipment and dies
2. Reduced delay, setup/adjustment time
3. More effective use of existing equipment
4. Precision control of equipment and dies
5. Conservation of materials and energy
6. Training and development of personnel

Goals

	Currently 1/78		Goals 3/81
1. Reduce failures	938/mo.	→	less than 50/mo.
2. Reduce equipment failure rate	1.03 cases/100 hrs.	→	less than 0.2 cases/100 hrs.
3. Reduce incidence of serious equipment failure	1.59%	→	less than 0.4%
4. Reduce non-operating time	5,800 hrs./mo.	→	under 2,900 hrs./mo. (reduced 50%)
5. Increase equipment operating rate	88.8%	→	more than 95%
6. Increase productivity	113%	→	more than 141% (25% increase)
7. Reduce defects in process	0.7%	→	less than 0.35%
8. Conserve energy	100% (1977)	→	less than 77% used (1980)
9. Increase number of improvements suggested	2.1/yr. per person (1977)	→	10/yr. per person (1980)
10. Reduce accidents	14.05 cases/1 million hrs.	→	7 cases/1 million hrs.
11. Increase safety rate	0.80 days/1,000 hrs.	→	0.03 days/1,000 hrs.

8/81
Results of TPM activities will be evaluated by the PM Prize Committee (when company is considered for the prize).

Figure 14. Example of TPM Basic Policies and Goals

STEP 5: FORMULATE A MASTER PLAN FOR
TPM DEVELOPMENT

The next responsibility of the TPM promotional head-quarters is to establish a master plan for TPM development. The daily schedule for promotion of TPM, beginning with the preparation stage before implementation, must be included.

Figure 15 is an actual TPM master plan taken from Central Motor Wheel Co., where TPM development centered on the following five basic improvement activities:

1. Improving equipment effectiveness through elimination of the six big losses (carried out by project teams)
2. Establishing an autonomous (operator) maintenance program (following a seven-step method)
3. Quality assurance
4. Establishing a schedule for planned maintenance by the maintenance department
5. Education and training to increase skills

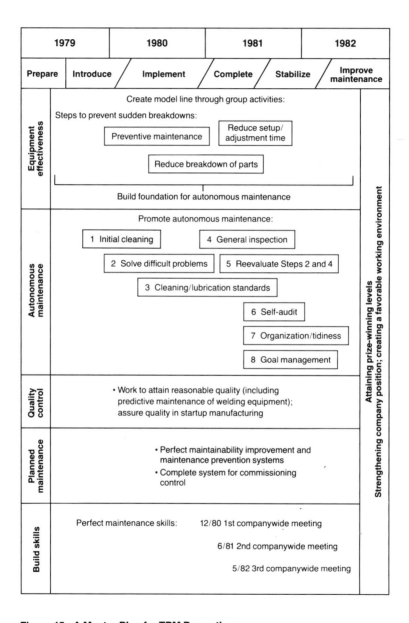

Figure 15. A Master Plan for TPM Promotion

5

TPM Implementation and Stabilization

In Chapter 4 we reviewed the first five steps in the TPM development program, which prepare for TPM implementation. Chapter 5 reviews the steps from TPM implementation to stabilization.

STEP 6: HOLD TPM "KICKOFF"

The TPM "kickoff" is the first step in implementation, the beginning of the battle against the six big equipment losses. During the preparatory stage (steps 1-5), management and professional staff play the dominant role. From that point on, however, the individual workers must move away from their traditional daily work routines and begin to practice TPM. Each worker now plays a crucial role. As someone once said, "There is no room for sitting in TPM," meaning that every person is a participant — there are no onlookers. For this reason, every worker must support top management's TPM policy through activities to eliminate the six big equipment losses.

The kickoff should help cultivate an atmosphere that increases worker morale and dedication. In Japan, this often takes the form of a meeting for all employees. Frequently representatives of client companies as well as affiliates and subcontractors are also invited. At the meeting, top managers report on the plans developed and work accomplished during the preparation phase, such as the TPM promotional structure, the basic TPM policy and goals, and the master plan for TPM development. Then, a representative of the workers affirms their common commitment to achieve the goals and win the PM Prize.

STEP 7: IMPROVE EQUIPMENT EFFECTIVENESS

TPM is implemented through five basic TPM development activities described in Chap.3, the first of which is to improve the effectiveness of each piece of equipment experiencing a loss.

Engineering and maintenance staff, line supervisors, and small group members are organized into project teams that will make improvements to eliminate losses. These improvements will produce positive results within the company. During the early stages of implementation, however, there will be those who doubt TPM's potential to produce results — even when they are shown how use of TPM in other companies increased productivity and quality, reduced costs, improved business results, and created a favorable working environment.

To overcome this doubt and build confidence in TPM, we demonstrate its effectiveness by focusing team efforts on equipment suffering from chronic losses during operation — pieces that will show marked improvement under an intensive three-month overhaul. Several pieces of equipment in each workshop are selected as models, and a project team is assigned to each piece.

These projects have a dual benefit: They prove the effectiveness of TPM and give the engineering and maintenance staff hands-on experience. Moreover, group leaders can use the experience gained to improve other equipment in the individual work centers.

Managers should not hesitate to apply IE, QC, or any other techniques to promote improvement in the target areas. *PM analysis* (developed by JIPM consultant Kunio Shirose) is another effective technique for eliminating chronic losses in equipment.

The "P" of PM analysis stands for the words "problem" or "phenomenon" and "physical," while the "M" stands for "mechanism," "machine," "man," and "material."

PM analysis consists of the following:

1. *Define the problem.* Examine the problem (loss) carefully; compare its symptoms, conditions, affected parts, and equipment with those of similar cases.
2. *Do a physical analysis of the problem.* A physical analysis clarifies ambiguous details and consequences. All losses can be explained by simple physical laws. For example, if scratches are frequently produced in a process, friction or contact between two objects should be suspected. (Of the two objects, scratches will appear in the object with the weaker resistance.) Thus, by examining the points of contact, specific problem areas and contributing factors are revealed.
3. *Isolate every condition that might cause the problem.* A physical analysis of breakdown phenomena reveals the principles that control their occurrence and uncovers the conditions that produce them. Explore all possible causes.
4. *Evaluate equipment, material, and methods.* Consider each condition identified in relation to the equipment,

jigs and tools, material, and operating methods involved and draw up a list of factors that influence the conditions.

5. *Plan the investigation.* Carefully plan the scope and direction of investigation for each factor. Decide what to measure and how to measure it and select the datum plane.

6. *Investigate malfunctions.* All items planned in step 5 must be thoroughly investigated. Keep in mind optimal conditions to be achieved and the influence of slight defects. Avoid the traditional factor analysis approach; do not ignore malfunctions that might otherwise be considered harmless.

7. *Formulate improvement plans.*

Table 9 is based on examples of PM analysis applied to chronic quality defects and minor stoppages.

STEP 8: ESTABLISH AN AUTONOMOUS MAINTENANCE PROGRAM FOR OPERATORS

The second of the five TPM developmental activities, *autonomous maintenance*, is the eighth step in the development program. It should be tackled right after the TPM kickoff.

Autonomous maintenance by operators is a unique feature of TPM; organizing it is central to TPM promotion within the company. The longer a company has been organized, the harder it is to implement autonomous maintenance, because operators and maintenance personnel find it difficult to let go of the concept: "I operate — you fix." Operators are used to devoting themselves full-time to manufacturing, and maintenance personnel expect to assume full responsibility for maintenance.

Such attitudes and expectations cannot be changed overnight, which is one of the reasons why it typically takes two to three years to progress from the introduction of TPM to its full implementation. Changing the thinking and the environment within a company takes time.

In promoting TPM, everyone from top to bottom in the organization must believe that it is feasible for operators to perform autonomous maintenance and that individuals should be responsible for their own equipment. In addition, each operator must be trained in the skills necessary to perform autonomous maintenance.

Some Japanese companies that have not yet implemented TPM insist that their operators perform autonomous maintenance activities such as inspection, lubrication, and cleaning. In most cases, however, the operators just go through the motions without actually making any effort. The daily check sheets they fill out reveal their attitude: Some operators check off items in advance (so as not to bother with it the next day); sometimes important tasks are neglected (for example, an oiler to be refilled regularly is unaccountably empty). Furthermore, since equipment is not properly maintained in these companies, abrasion, shaking, loosening, contamination, and corrosion cause breakdowns and quality defects.

Autonomous Maintenance in Seven Steps

The 5 S's: *seiri, seiton, seiso, seiketsu,* and *shitsuke* (roughly, organization, tidiness, purity, cleanliness, and discipline) are basic principles of operations management. At present, though most factories apply some of these principles, many do so rather superficially. Management is often more concerned about appearances, such as painting the factory interior or equipment, and neglects internal cleaning of revolving and moving parts. Covering up dirt, dust, and rust with paint is like wearing heavy make up to cover unhealthy skin. The make up soon cracks and smears revealing the unhealthy skin underneath. Covering it up may actually worsen the condition.

Factory	Phenomenon	Description	Basic conditions	Relevance of equipment, materials, jigs, and tools
Extrusion process of vinyl chloride	Pyrolysis	Carbonization caused by excessive heat; carbonization accompanied by partial clogging caused by abnormal flow	1. Gap between cylinder and screw	• Cylinder abrasion
			2. Causes in the screw	• Eccentricity of the screw • Scratches on the screw • Screw abrasion • Dirt on the screw
			3. Assembly precision of individual parts	• Assembly precision
			4. Precision of individual parts	• Dirt in surrounding area

Table 9. Example of PM Analysis (1)

Factory	Phenomenon	Description	Basic conditions	Relevance of equipment, materials, jigs, and tools
Dry battery process	Batteries falling on revolving table	Loss of balance accompanying shift of center of gravity caused by external conditions (Shock, friction, shaking, etc.)	1. Conditions creating friction • Contact between revolving table and product • Factors inherent in product (warping of bottom, abnormal attachment)	Omitted
			2. Conditions creating shaking • Factors inherent in revolving table (undulating, involuntary shaking) • Contact between revolving table and surrounding guides	• Table surface conditions, levelness, and shaking • Irregular revolving • Guide shape, position, angle and surface conditions • Contact point conditions

Table 9. Example of PM Analysis (2)

JIPM recommends that companies wishing to avoid superficial autonomous maintenance adopt a seven-step approach that includes progressive mastery of each of the 5 S's (Table 10). Individual workers acquire the skills corresponding to each step through training and practice. Only after training in one skill has been completed and confirmed is the worker permitted to progress to the next step.

Table 10 outlines the seven-step method developed by JIPM consultant Fumio Goto. (For further details, refer to Chapter Four of *TPM Development Program* [Cambridge: Productivity Press, forthcoming]). Here the goals and targets of each step are summarized.

1. Initial cleaning. Operators develop an interest in and concern for their machines through cleaning them thoroughly. Cleaning is an educational process that raises some questions ("why does this part accumulate dirt so quickly?") and answers others ("there's no vibration when this bolt is properly secured"). Operators learn that cleaning *is* inspection. They also learn basic lubricating and bolting techniques and become skilled in detecting equipment problems.

2. Countermeasures for the causes and effects of dirt and dust. The more difficult it is for the individual to perform initial cleaning, the stronger will be his/her desire to keep the equipment clean and, thus, to reduce cleaning time. Measures to eliminate the causes of dust, dirt, chips, and so on, or to limit scattering and adherence (*e.g.*, by using covers and shields) must be adopted. If a cause cannot be removed completely, more efficient cleaning and inspection procedures must be designed for the problem areas. Each work shop is responsible for cleaning and improving its work area, but engineering and maintenance staff must cooperate with them and support their efforts.

Step	Activities
1. Initial cleaning	Clean to eliminate dust and dirt mainly on the body of the equipment; lubricate and tighten; discover problems and correct them
2. Countermeasures at the source of problems	Prevent cause of dust, dirt and scattering; improve parts that are hard to clean and lubricate; reduce time required for cleaning and lubricating
3. Cleaning and lubrication standards	Establish standards that reduce time spent cleaning, lubricating, and tightening (specify daily and periodic tasks)
4. General inspection	Instruction follows the inspection manual; circle members discover and correct minor equipment defects
5. Autonomous inspection	Develop and use autonomous inspection checksheet
6. Organization and tidiness	Standardize individual workplace control categories; thoroughly systemize maintenance control • Inspection standards for cleaning and lubricating • Cleaning and lubricating standards in the workplace • Standards for recording data • Standards for parts and tools maintenance
7. Full autonomous maintenance	Develop company policy and goals further; increase regularity of improvement activities Record MTBF analysis results and design counter-measures accordingly

Table 10. Example of the Seven Steps for Developing Autonomous Maintenance

3. Cleaning and lubricating standards. In steps 1 and 2, operators identify the basic conditions that should apply to their equipment. When this has been done, the TPM circles can set standards for speedy and effective basic maintenance work to prevent deterioration, *e.g.*, cleaning, lubricating, and bolting for each piece of equipment.

Obviously, the time available for cleaning, lubricating, bolting, and detecting minor defects is limited. Supervisors must give operators reasonable targets for time to be spent on cleaning and lubrication — for example, ten minutes a day before and after operation, thirty minutes on the weekends and one hour at the end of the month.

If the standards set by the operators cannot be maintained within the target times, they must improve cleaning and lubricating practices. This can be accomplished by investigating innovative ideas, such as visual controls to show limits on oiler level gauges, along with better positioning of oilers and more efficient lubricating methods. In such cases, operators can make changes with the full support and cooperation of supervisors and staff.

Figure 16 shows an example of the cleaning and lubricating standards tentatively set by one circle. They are later reevaluated as autonomous maintenance standards in Step 5, *Autonomous Inspection.*

4. General inspection. Steps 1 through 3 are carried out to prevent deterioration and control the basic conditions of equipment maintenance — cleaning, lubricating, and bolting. In Step 4, we attempt to measure deterioration with a general inspection of equipment. Moreover, in working to restore the equipment to good operating conditions, the operators' equipment competence is increased.

Initially, TPM circle leaders are trained in these inspection procedures (one inspection category at a time) using a general inspection manual prepared by supervisors and staff. These leaders then share what they have learned with their circle members. Team members work together to target problem areas discovered during the general equipment inspection. Finally, with the assistance of staff and maintenance personnel, the circle takes action to correct deterioration and improve the affected areas.

Figure 17 shows the developmental steps needed to train workers and implement a general equipment inspection. Figure

18 illustrates examples of inspection training (by category) and a typical training schedule.

General inspection training must be carried out one category at a time, beginning with skill development. Its effectiveness is audited and reinforced with additional training and practical applications. This cycle of training, application, auditing, and modification is repeated for each inspection category.

This fourth step may take a long time to complete, because all operators must develop the ability to detect abnormalities. It is the best method to produce competent operators, however, so this step should not be rushed. Positive results cannot be achieved until each worker acquires all the necessary skills.

The first three steps of autonomous maintenance focus on meeting basic requirements, therefore efforts at this early stage may not always show dramatic results. By the end of Step 4, however, the company should see amazing changes, such as an 80 percent reduction in equipment failures or an overall equipment effectiveness rate of over 80 percent.

If results have not appeared by this time, the skills taught in the early steps probably have not been mastered. It may also indicate a generally low level of technical expertise. If this is the case, it is better to start over and begin by working to raise the technical level.

5. Autonomous inspection. In Step 5, the cleaning and lubricating standards established in Steps 1 through 3 and the tentative inspection standards are compared and reevaluated to eliminate any inconsistencies and to make sure the maintenance activities fit within the established time frames and goals.

By the time operators are thoroughly trained to conduct the general inspection (step 4), the maintenance department should set up an annual maintenance calendar and prepare its own maintenance standards. Standards developed by workshop circles must then be compared to these maintenance standards to correct omissions and eliminate overlapping in

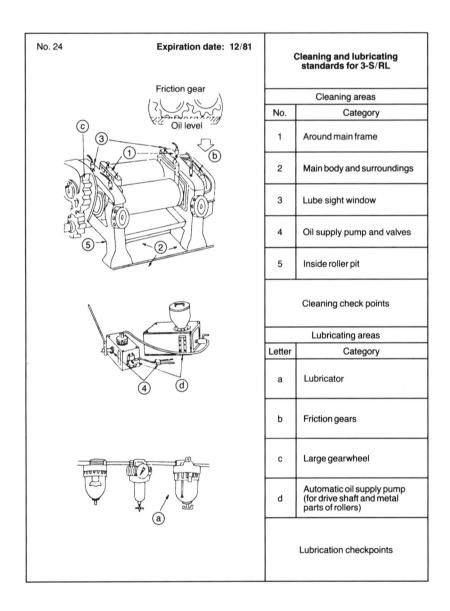

No. 24	Expiration date: 12/81	Cleaning and lubricating standards for 3-S/RL	

Cleaning areas

No.	Category
1	Around main frame
2	Main body and surroundings
3	Lube sight window
4	Oil supply pump and valves
5	Inside roller pit

Cleaning check points

Lubricating areas

Letter	Category
a	Lubricator
b	Friction gears
c	Large gearwheel
d	Automatic oil supply pump (for drive shaft and metal parts of rollers)

Lubrication checkpoints

Friction gear
Oil level

Figure 16. Example of Cleaning and Lubricating Standards (Tōkai Rubber Industries)

Cleaning and lubricating standards for 3-S/RL	Plant manager: Section chief: PM engineer: Foreman:			Somura plant		
Cleaning standards	Cleaning methods	Cleaning tools	Cleaning time	Cleaning cycle		
				Day	Wk.	Mo.
No rubber scrap adhering to frame	Remove with steel scraper; sweep up		15 min.	○		
No scattering of rubber scrap	Sweep away with broom		5 min.	○		
Oil level easy to check	Wipe clean with cotton waste		3 min.	○		
No oil and dirt	Wipe clean with cotton waste		10 min.	○		
Not leaking or dirty			30 min.			○
1. Tighten the automatic supply pump ring joint bolts						
2. Tighten oil supply valve and check for leaks						
3. Tighten the stock guide fixing bolts						
Lubrication standards	Lubrication methods	Lubrication equipment	Lubrication time	Lubrication cycle		
				Day	Wk.	Mo.
Oil level must be between upper and lower limit (#32)	Pour by hand		10 min.	○		
Oil level half-way up gear teeth (#220)	Oil can		5 min.			○
Gearwheel well-lubricated (open gear/oil)	Drip in through oil supply port with spatula		5 min.			○
Adequate oil (as measured by the oil gauge) (R50)	Use oil applicator		3 min.	○		
1. Secure large gearwheel cover (no rattling)						○
2. Secure attachment bolts for automatic oil supply pump						○
3. Check 3-unit FRL (filter, regulator, lubricator) and drip rate					○	

Figure 16. Example of Cleaning and Lubricating Standards (Tōkai Gomu Kōgyō)

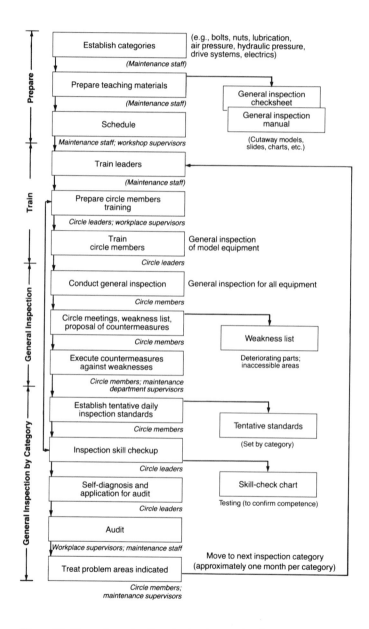

Figure 17. Procedures for Developing Inspection Education and Training

General inspection categories	
Air pressure 1	Piping, air regulator (filter, regulator, lubricator)
Air pressure 2	Compressed air valves and cylinders
Lubrication	Function and types of lubricants
Basic operations	Correct tightening of bolts and nuts
Electrics	Limit switches; proximity switches
Drive systems, moving parts	Motor; reduction gears; transmissions; sprockets; V-pulleys; V-belts
Hydraulics	Hydraulic valves, cylinders, and fluids

Equipment-specific categories

- Leader education (by TPM promotion office)
  1. Function, structure and names of parts
  2. Problems and counter-measures
  3. Focus of inspection, methods, standards, etc.
  4. Inspection practice and evaluation

- Operator education (by group leaders and trainers)
  1. Function, structure and names of parts
  2. Problems and counter-measures
  3. Inspection training and evaluation (OJT, meetings)
  4. Self-inspection and evaluation (OJT, meetings)

- Topics

  Air pressure 1, Air pressure 2, lubrication, basic operations, electrics, drive systems, hydraulics, one category per month

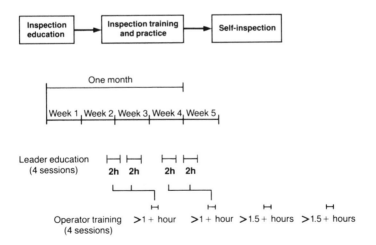

Figure 18. Scheduling Pattern for General Inspection Education and Training

individual categories. The responsibilities of the two groups must be clearly defined so that a complete inspection is carried out for each category.

6. Organization and tidiness. *Seiri*, or organization, means to identify aspects of the workplace to be managed and set appropriate standards for them. This is a job for managers and supervisors, who must minimize and simplify the objects or conditions to be managed. *Seiton*, or tidiness, which means adhering to established standards, is mainly the operators' responsibility. Part of their circle activities should always focus on improvements that make standards easier to follow.

Seiri and *seiton* are thus improvement activities that promote simplification, organization, and adherence to standards — ways of ensuring that standardization and visual controls are instituted throughout the factory.

Steps 1 through 5 emphasize activities concerned with the inspection and maintenance of basic equipment conditions (cleaning, lubricating, and bolting). The role of the operator is much broader than this, however.

In Step 6, supervisors and managers take the lead in completing implementation of autonomous maintenance by evaluating the role of operators and clarifying their responsibilities. What should operators do to prevent losses from breakdowns and defects, for example, and what additional skills should they possess? On the basis of operators' experience up to this point, managers may want to expand the scope of their equipment-related activities.

In addition to maintaining basic conditions and inspecting equipment, operators may also be responsible for:

- Correct operation and setup (setting conditions and checking product quality)

- Detection and treatment of abnormal conditions
- Recording data on operation, quality, and processing conditions
- Minor servicing of machines, molds, jigs, and tools

Table 11 is an example of standards for organization and tidiness. In it, step 6 has been further divided into six sub-steps, which are described in detail.

Focus	Elements
Operator's responsibility	Organize standards for operator responsibilities; adhere to them faithfully (including data recording)
Work	Promote organized and orderly operations as well as visual control of work-in-process, products, defects, waste, and consumables (such as paint)
Dies, jigs, and tools	Keep dies, jigs, and tools organized and easy to find through visual control; establish standards for precision and repair
Measuring instruments and fool-proof devices	Inventory measuring instruments and fool-proof devices and make sure they function properly; inspect and correct deterioration; set standards for inspection
Equipment precision	Operators must check precision of equipment (as it influences quality) and standardize procedures
Operation and treatment of abnormalities	Establish and monitor operation, setup/adjustment, and processing conditions; standardize quality checks; improve problem-solving skills

Table 11. Example of Organization and Tidiness Standards in Autonomous Maintenance

7. Full implementation of autonomous maintenance. Through circle activities lead by supervisors (through step 6), workers develop greater morale and competence. Ultimately, they become independent, skilled, and confident workers who can be expected to monitor their own work and implement improvements autonomously.

At this stage, circle activities should focus on eliminating the six losses and implementing in each workshop the improvements adopted for the model equipment by the project teams.

Autonomous Maintenance Auditing

Audits of circle activities and equipment by supervisors and staff play an important role in the successful development of an autonomous maintenance system. To conduct them effectively, supervisors and staff must understand the workplace environment thoroughly; they must provide the circles with appropriate instructions and encouragement to give workers a sense of accomplishment as they complete each step. Figure 19 illustrates the cycle of autonomous maintenance auditing.

Step 9: SET UP A SCHEDULED MAINTENANCE PROGRAM FOR THE MAINTENANCE DEPARTMENT

The ninth step in the development program is also one of the five basic TPM activities described in chapter 3 — a scheduled or periodic maintenance program for the maintenance department.

As we mentioned earlier, scheduled maintenance carried out by the maintenance department must be coordinated with the autonomous maintenance activities of the operations department so the two departments can function together like the wheels of a car.

Until general inspection becomes part of workers' routine, the assistance of the maintenance department will be required more often than it was before the TPM development program was introduced. For example, operations will rely on maintenance to point out weaknesses and design countermeasures for problem areas. Moreover, accidental breakdowns, while gradually decreasing, will continue to demand attention. Thus, the workload of the maintenance department will probably hit an all time high. This temporary excess work load should be handled promptly through overtime and subcontracting, however, to support the commitment of the operators. Otherwise, the operators will lose their enthusiasm for developing autonomous maintenance.

The volume of maintenance work will diminish once again when general inspection has become part of the operators' routine. The number of breakdowns will decrease significantly and maintenance activities will also lessen. At this point, the maintenance department should focus on its own organization.

Development of a scheduled or periodic maintenance program should actually begin before the operators' general inspection procedure (Step 4) has been completely set up. As mentioned earlier, the maintenance department must develop equipment standards independently, so that during the autonomous inspection stage (Step 5) they can be compared against the standards being set by the operations department. A clear division of the responsibilities of the two departments is the key to thorough and effective inspection and can be accomplished only when both sets of standards are combined.

Since productive maintenance was first introduced, the subject of scheduled maintenance has been widely discussed and does not need to be covered again. All that needs to be

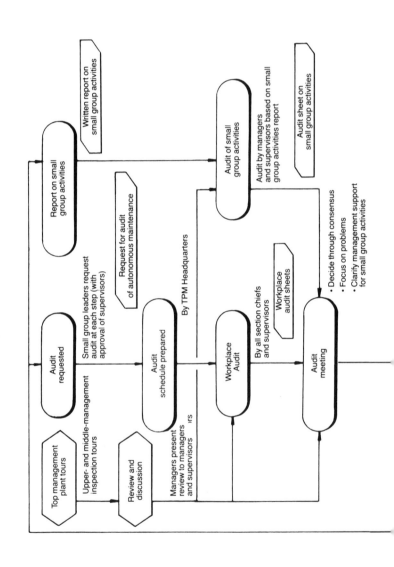

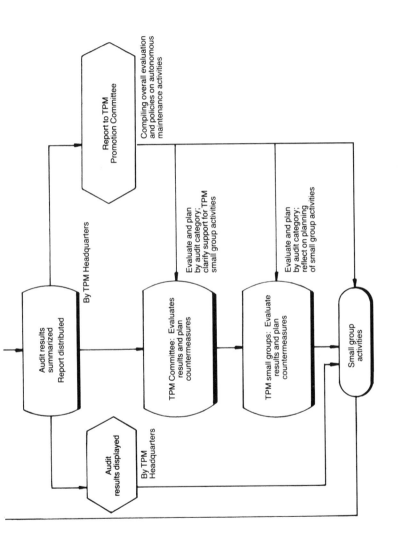

Figure 19. Autonomous Maintenance Audit Cycle

said here is that if scheduled maintenance is inadequate in a company, it should be reevaluated and improved as part of the TPM development program.

Furthermore, to maximize the effectiveness of its activities, the maintenance department should reevaluate control of spare parts, dies, tools, inspection devices, and drawings. Figures 20, 21, and 22 present examples of scheduled maintenance from the 1982 PM Prize winner Central Motor Wheel Co. Figure 20 is the company's productive maintenance organization schedule, and Figure 21 is a flow chart of its maintenance management. Maintenance on the production line is performed during downtime scheduled for regular weekday operations.

A distinctive feature of this maintenance system is the "L & S maintenance meeting" held daily at 9 a.m. At these meetings production line managers and supervisors (L) and members of the staff (S), such as maintenance and engineering supervisors, discuss planning and scheduling for production line stops and maintenance work. The meetings promote speedy implementation of monthly and weekly scheduled maintenance and more efficient handling of daily breakdowns.

Figure 22 illustrates the computerized maintenance planning and management system at Fuji Film's Yoshida Minami plant. This system has been used effectively for overall scheduling as well as for semi-quarterly and monthly equipment inspection and parts procurement scheduling.

STEP 10: CONDUCT TRAINING TO IMPROVE OPERATION AND MAINTENANCE SKILLS

Improving operational and maintenance skills is the fourth TPM development activity and the tenth step of the TPM development program.

In Japan, the large steel and electronics corporations provide their employees with technical training at well-equipped

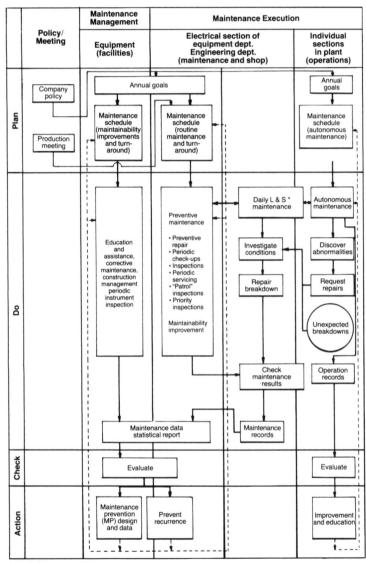

* L & S: "line and staff."

Figure 20. Example of Productive Maintenance System

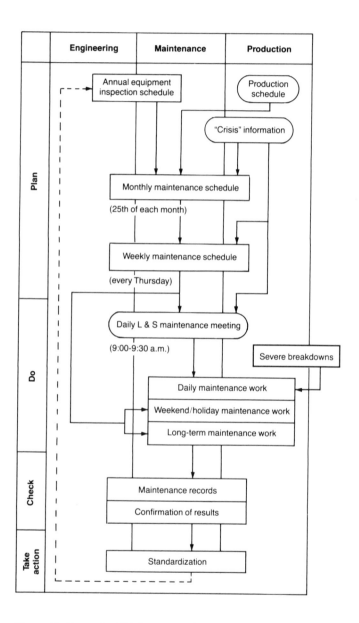

Figure 21. Example of Maintenance Control Flowchart

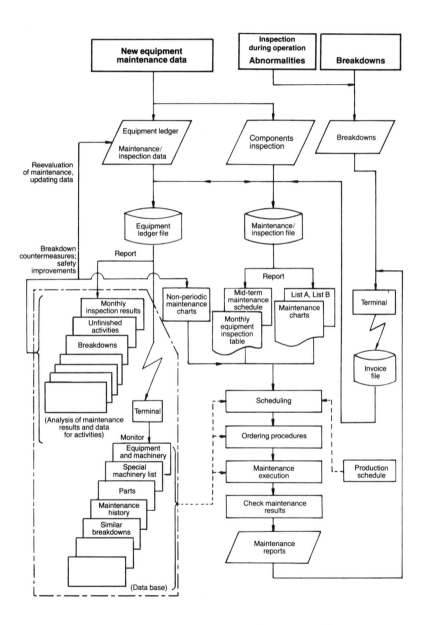

Figure 22. Example of Maintenance Activities Based on EDPS

centers, but many other Japanese companies underestimate the value of training, especially training in maintenance techniques. Education and training are investments in people that yield multiple returns. A company implementing TPM must invest in the training that will enable employees to manage their equipment properly. In addition to training in maintenance techniques, operators must also sharpen their conventional operation skills.

The content and organization of introductory materials may not vary much from company to company; technical education and training for operation and maintenance, on the other hand, must be tailored to the individual requirements of the workplace. At the Mizushima plant of Nihon Zeon, winner of the 1982 PM Prize, for example, operators were taught to perform daily inspections and simple repairs. Using equipment that simulated abnormal conditions in the production plant, operators learned first-hand how to deal with unusual or crisis situations. This simulation training was conducted at different levels (see Figure 23).

Maintenance personnel are like doctors — they must be competent, otherwise their patient's condition can only worsen. To promote quality in equipment maintenance, technical qualifying examinations for equipment maintenance personnel were established in Japan in 1984. These examinations are similar to the national examinations for mechanics. In the past, individuals were certified as equipment maintenance personnel when they qualified in such areas as finishing and machining. Now, however, maintenance workers can be certified in their own field. In addition, the Ministry of Labor has established a qualifying system for electrical maintenance.

Responding to the demand for maintenance training, JIPM, with the help of its member companies, has established

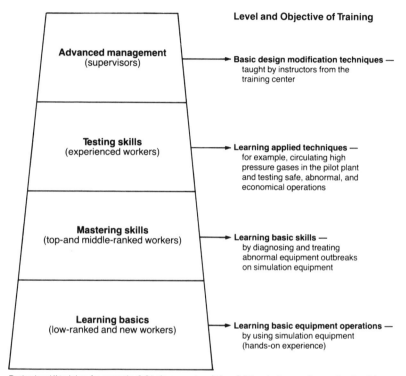

During level II training, for example, 6-8 trainees are taught by a full-time lecturer and supervisor for eight full days, combining study, discussion and practical application

Figure 23. Example of Simulation Training for Operators

training centers at most of its branch offices. In the four-month, four-part technical course taught at these centers, maintenance workers from companies in the process of implementing TPM learn more about their equipment. Table 12 summarizes the syllabus for this course.

Required days	Subject	Elements
3 days	1. Nuts and bolts	1. Basics of connecting nuts and bolts 2. How to avoid loosening 3. Handling maintenance tools and equipment
3 days	2. Key matching	1. Types and appropriate use of keys 2. Filing and matching keys 3. Techniques for withdrawing keys
3 days	3. Shaft and bearing maintenance	1. Fitting shafts and bosses 2. Attaching bearing and performing test runs 3. Shaft-case lubrication and related problems
3 days	4. Transport equipment	1. Gear-driven unit basics 2. Chain-driven unit basics 3. Belt and brake system basics
3 days	5. Sealing methods	1. Importance and basic techniques of sealing 2. Types of gaskets 3. Assembling O-rings and taper pipe thread and performing pressure tests

Table 12. Curriculum for the Basic Equipment Maintenance Technical Training Course

STEP 11: DEVELOP EARLY EQUIPMENT MANAGEMENT PROGRAM

The last category of TPM development activities is early equipment management.

When new equipment is installed, problems often show up during test-running, commissioning, and startup even though design, fabrication, and installation appear to have gone smoothly. Engineering and maintenance engineers may have to make many improvements before normal operation can begin. Even then, startup period repairs, inspection, adjustment, and the initial lubrication and cleaning needed to prevent deterioration and breakdown are often so difficult to carry out that supervising engineers become thoroughly discouraged. As a result, inspection, lubrication, and cleaning may be neglected, which needlessly prolongs equipment downtime for even minor breakdowns.

These startup troubles and the equipment improvements implemented subsequently often represent the tying up of loose ends from the design and construction stages. This phenomenon does not have to result from increasing the scale, speed, and automation of production to match advances in technology; often it can be avoided by building the appropriate processing and operating conditions into the equipment.

The quality of subsequent production maintenance is determined largely by whether the technology for ensuring equipment reliability and maintainability was developed by engineering, design, and maintenance staff out of their direct experience and concerted efforts, or whether it was simply brought in from outside the company.

Early equipment management is performed mainly by production engineering and maintenance personnel as part of a comprehensive approach to maintenance prevention (MP) and maintenance-free design. These goals are promoted through improvement activities at various stages: the equipment investment planning stage, design, fabrication, installation and test running, as well as commissioning (when normal operation with an actual flow of products has been established). Debugging (detecting and correcting errors and faults) is included in these activities.

These activities are aimed at:

- Achieving the highest levels possible within the limits established at the equipment-investment planning stage
- Reducing the period from design to stable operation
- Progressing through this period efficiently, with minimum labor and no workload imbalance
- Ensuring that the equipment designed is at the highest levels of reliability, maintainability, economical operability, and safety

By working together with design engineers during commissioning to eliminate problems at the source and by promoting activities within individual project teams, the engineering and maintenance staff can absorb and apply knowledge about MP (maintenance prevention) design.

Commissioning is the stage of actual production after equipment has been installed and test runs are completed; it is a time for debugging the equipment for speedy progress to stable operation. It is essential during earlier stages to prevent faults from carrying over into startup, however, commissioning is the last opportunity for detecting and correcting faults that were not anticipated. Frequent failures during this stage may indicate that earlier opportunities for improvement were neglected.

The goal of TPM is to maximize equipment effectiveness, in other words, to pursue economic life cycle cost (LCC). According to B.S. Blanchard (*Design and Manage to Life Cycle Cost*, [Forest Grove, Oregon: M/A Press, 1978]), virtually 95 percent of LCC is determined at the design stage (see Figure 24). Certainly, maintenance and energy costs of operation are determined by the equipment's original design. Efforts to reduce LCC after the design stage will affect only five percent of the overall figure.

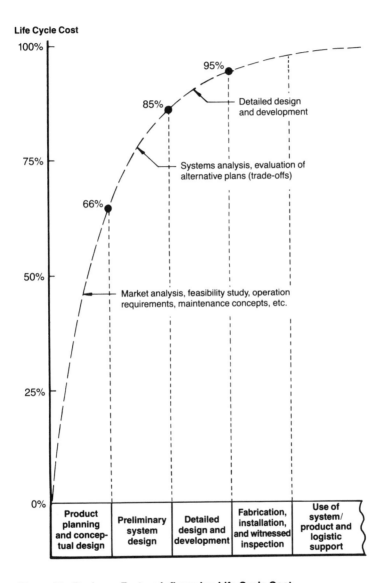

Figure 24. Business Factors Influencing Life Cycle Cost

The following improvement activities can have a positive impact on life cycle cost:

- Economic evaluation at the equipment-investment stage
- Consideration of MP or maintenance-free design and economic LCC
- Effective use of accumulated MP data
- Commissioning control activities
- Thorough efforts to maximize reliability and maintainability

It is essential, therefore, that equipment engineers study situations where design has been maximized for reliability and maintainability. This information should then be evaluated and developed into design technology standards.

In many cases, unfortunately, poor horizontal communication between equipment planning, operations, and maintenance departments precludes the use of technical improvement data obtained from routine PM activities in reliability and maintenance design. Maintenance engineers do not share data relating to maintainability and reliability that could be relevant at the design and fabrication stages; and design engineers do not standardize general technical data or use the maintenance data they receive. When maintenance and design engineers cooperate to close the gap between maintenance and design technology, much waste can be avoided.

Figure 25 presents a step-by-step problem-prevention control chart used by project teams to promote early equipment management at Tokai Rubber Industries, a 1981 PM Prize winner. The chart was developed to improve horizontal communication between related departments.

This method anticipates potential problems at each stage of early equipment management (*e.g.,* planning, design, blueprint, fabrication, and witnessed inspections). Countermeasures are developed and applied before problems actually occur and are followed by checkups to confirm their effectiveness.

The problem-prevention control chart is initially time-consuming to use, but practice makes the process go faster and more easily. Taking this extra time in the early stages greatly reduces the number of problems occurring later and helps establish an efficient management program. As Figure 26 shows, it drastically reduced the number of startup failures at Tokai Rubber Industries.

STEP 12: IMPLEMENT TPM FULLY AND AIM FOR HIGHER GOALS

The final step in the TPM development program is to perfect TPM implementation and set even higher goals for the future. During this period of stabilization everyone works continuously to improve TPM results, so it can be expected to last for some time.

At this point, Japanese companies are evaluated for the PM Prize. Even after a company receives the PM Prize, however, its drive for improvement must continue — winning the PM Prize simply symbolizes a new beginning. According to a top-ranking manager at a PM prize-winning workplace, "This award does not mark our completion of TPM, but it does signify that we started out on the right foot. This will enable us to try even harder."

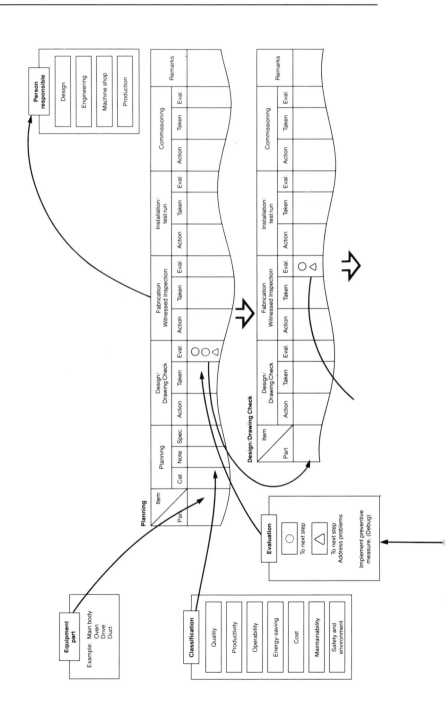

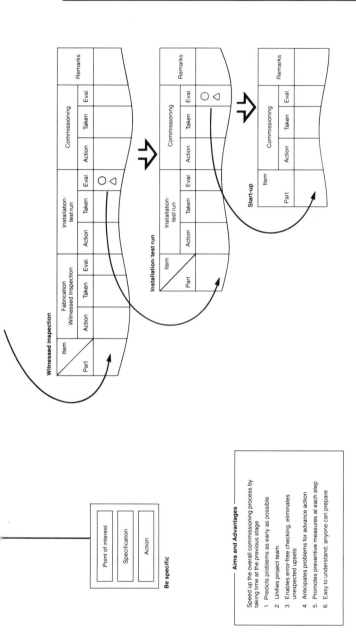

Figure 25. Step by Step Maintenance Testing Control Chart

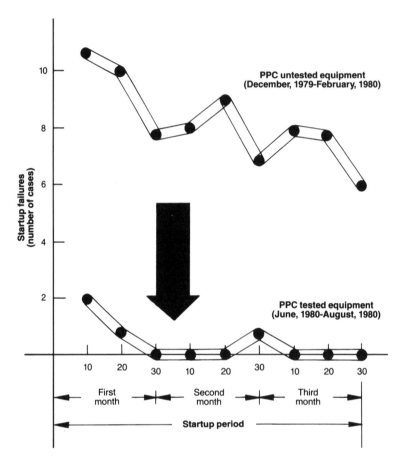

Note: PPC is an abbreviation for problem-prevention control.
Actual example based on comparison of similar types of equipment.

Figure 26. Reduction in Startup Failures (Tōkai Rubber Industries)

6
TPM Small
Group Activities

The most outstanding feature of TPM small group activity is its structure of overlapping groups (introduced in Chapter 4) integrating organizational and small group improvement activity. Japanese small groups have attracted attention worldwide and many countries, including the United States, are gradually implementing them.

INTEGRATING SMALL GROUP ACTIVITIES INTO THE ORGANIZATIONAL STRUCTURE

Japanese-style small group activities began with the quality control circle, introduced in 1962. The American concept of zero defects (ZD), which is an individual rather than a group activity, became popular three years later. NEC, the first Japanese firm to implement it, combined this individual improvement activity with the Japanese-style QC circle to form ZD group activities.

Later, the Japanese steel industry followed suit with the widely used "JK" (*jishu kanri* or "autonomous management") activities. Since then, many other companies have developed their own terminology and procedures for conducting QC circles and ZD groups. In fact, most Japanese companies now promote some form of small group activity, even service industries such as hotels, banking, and insurance.

In spite of differences in terminology and operating procedures, small groups can be divided into two broad categories, one originating in the early QC circle, the other in the ZD movement. These two groups are distinctive in several ways.

QC circles began as study groups to teach shop floor supervisors quality control techniques and evolved into problem-solving small groups for a larger segment of the worker population. Circles are organized by subject or theme to deal with specific problems within the larger TQC program. In terms of organizational theory, they are *informal organizations*. Formed by workers, they are independent of the existing organizational structure. Participation is voluntary.

ZD groups, on the other hand, were first used in the United States at Martin Marietta as a means of involving all employees in solving the problem of delayed delivery. They were imported by the Japanese and adapted to fit their small group activities. Japanese ZD groups participate in management-based activities to find solutions to company problems and work to achieve company goals.

Organizational theory considers Japanese ZD groups "formal organizations" because they are conducted within the existing organizational structure. For example, in the independent QC circles, leaders are typically selected by circle members. In the management-based ZD groups, on the other hand, shop floor supervisors most often assume leadership roles. Of course, when a supervisor has too many subordinates, subgroups (ranging from five to ten members) must be formed and additional leaders chosen.

Typically, independent QC circle activities are conducted outside of normal working hours, during "free time," (*e.g.*, breaks, after work hours, and weekends and holidays). Because circle activities are voluntary, employees at most Japanese companies do not receive overtime compensation for participation in them. By contrast, the formal ZD groups can meet during work hours under the supervisor's direction as well as during free time, and some companies pay overtime compensation for these outside activities.

The improvement themes selected and the goals set also reflect the differences between QC circle and ZD group activities. QC circles are formed around specific themes and goals are set within each theme. Once the goals are achieved, the QC circles are reorganized around new themes. Ideally, themes are selected independently of annual management goals. This is made possible by the informal, voluntary nature of the circles. Although companies seem to respect circle autonomy and allow the circles to choose their own themes, management is increasingly encouraging TQC activities as part of companywide improvement activities and promoting themes that support the achievement of annual goals.

ZD groups, on the other hand, *must* choose goals consistent with the annual company goals because, ultimately, ZD is aimed at the elimination of defects and promotes the attainment of all related goals. Group members independently discuss and set sub-goals, such as lower costs, shorter deadlines, and the introduction of new methods.

Although QC circles and ZD small groups differ organizationally, they have often merged and influenced each other, and their distinctive features have become blurred. Many corporations have used both types of groups to develop their own unique systems.

JIPM supports the use of the "autonomous small groups" advocated by Professor Emeritus Odaka of Tokyo University. According to Odaka, Japanese small group activities have flourished, even though their position within the organizational structure has remained ambiguous. He argues that it is time to integrate small groups into the corporate structure so that their activities can complement and enhance other organizational activities.

Accordingly, TPM small group activities are based on the ZD model and built into the organizational framework. Specifically, TPM promotes autonomous maintenance by operators through small group activities. In TPM, the typically management-directed activities of equipment cleaning, lubrication, bolting, inspection, etc., are performed as small group activities.

During the TPM implementation stage, the time spent on different small group activities is carefully monitored. Activities are categorized and recorded (*e.g.*, as maintenance operations, education and training, and meetings). Documenting how small group time is spent allows companies to compensate their employees properly. For example, during the early stages of autonomous maintenance, much time is spent on maintenance operations and education and training; later on, more time is spent in meetings. Operators should be paid for overtime when they perform maintenance operations after work hours; employees attending training programs after work hours should receive education compensation; and if a certain number of hours per month is allotted for meetings, then those exceeding the limit should be held after regular work hours.

By the time factory workers are able to conduct the general inspection (Step 4) of autonomous maintenance, they can enjoy a real sense of accomplishment when, for example, their efforts reduce breakdowns by as much as 80 percent, increase productivity, and make work easier.

This sense of accomplishment naturally enhances morale and motivation and finds expression in longer and more frequent meetings as well as a greater number of improvement suggestions from workers. Moreover, when maintenance personnel disassemble equipment on weekends for servicing, operators will want to participate in order to learn. At this point, overtime compensation will no longer be an issue.

To promote operators becoming more capable and well-trained, managerial staff should lead group activities through Step 6, orderliness and tidiness. Pursuing ever higher goals, workers should be able to carry out autonomous maintenance independently from Step 7 on.

SMALL GROUP GOALS COINCIDE WITH COMPANY GOALS

Why do we advocate integrating TPM small group activities into an organizational structure? Here is another perspective:

What does a "small group" do? According to authors Hirota and Ueda, in *Small Group Activities: Theory and Reality* (Tokyo: Japan Labor Research Group, 1975), it "promotes itself and satisfies company goals as well as individual employee needs through concrete activities."

Teams called "circles" or "groups" set goals compatible with the larger goals of the company and achieve them through group cooperation or teamwork. This enhances company business results and promotes activities that satisfy both individual employee needs (self-satisfaction, success, motivation) and the needs of the organization. TPM small group activities are representative of this type.

In his book *New Patterns of Management* (New York: McGraw-Hill Book Co., 1961), behavioral scientist Rensis Likert compared companies and factories with high productivity and those with low productivity. He studied the impact of

different management policies and levels of employee con-
sciousness and behavior on productivity.

Likert discovered that the high-producing companies
strove to improve product variables (business factors such as
profits and sales) as well as intermediate variables (namely
human resources, which serve as intermediaries for the business
results) (see Figure 27). These companies attempt to improve
both business results and working conditions. Low-producing
companies and factories, on the other hand, ignore the
human factor and focus solely on product variables. Likert
calls the former "participative" and the latter "authoritarian"
management.

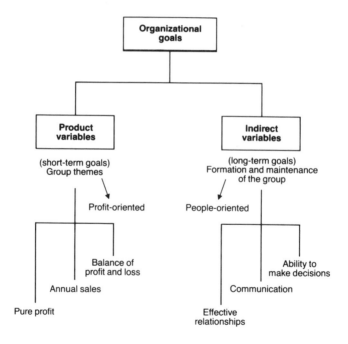

Figure 27. Organizational Goals

Likert argues that participative management is ideal because it encourages confidence among employees and promotes consistently high productivity. Authoritarian management, on the other hand, encourages submission based on fear among employees. Consequently, even if higher productivity can be achieved for a short time, low employee morale will eventually lead to a decline in productivity.

Small group activities in the factory should be based on participative management such as that advocated by Likert. Small group goals should be the same as company goals — to improve productivity and working conditions. In Japan, participative management is achieved through small group activities that have produced outstanding results. Ironically, American companies are only now beginning to study Japanese small group activities. Quality control developed in much the same way: Many American companies did not recognize the importance of the teachings of Juran, Deming, and Crosby until after Japan applied their methods and surpassed the U.S. in product quality.

EVALUATING THE MATURITY OF SMALL GROUP ACTIVITIES

If small group goals are the same as company goals, we can evaluate their progress by measuring the degree to which group activities contribute to the achievement of company goals. Progress in small group activities can be broken down into four stages:

Stage 1: Self-development. At first, group members must master techniques; their motivation increases as they recognize the importance of each individual.
Stage 2: Improvement activities. Group improvement activities are proposed and implemented, leading to a sense of accomplishment.

Stage 3: Problem-solving. At this stage, small group goals that complement company goals may be selected and the group becomes actively involved in problem-solving.
Stage 4: Autonomous management. The group selects high-level goals consistent with corporate policy and manages its work independently.

The small group activities in Stages 1 through 3 are not inconsistent with a traditional organization based on order and control, with newly-implemented small group activities taking place on the shop floor. During Stage 4, however, new human resource-oriented organizations are based on the self-managed small group model and are highly motivated to achieve company goals. Thus, during the final stage, true participative management is established. This is the goal of TPM small group activities.

THE FUNCTION OF TOP MANAGEMENT IN SMALL GROUP ACTIVITIES

Many experts say that the keys to success in small group activities lie in three conditions: motivation, ability, and a favorable work environment. Management is responsible for actively promoting these three conditions.

Of these three keys, motivation and ability are the workers' responsibility, but the creation of a favorable work environment is outside their control. This environment has both physical and psychological components that must be satisfied.

Figure 28 shows the division of these three conditions into two subgroups: human and environmental problems.

Management's first responsibility is to provide the special training necessary to develop a workforce of capable, motivated, and truly autonomous workers who possess the skills

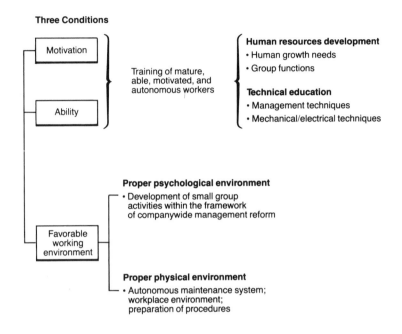

Figure 28. Key to Success of Small Group Activities

to perform autonomous maintenance. Human education, which explores "what human beings are," as well as technical training in maintenance and operational techniques must be provided.

Education is the source of motivation because it enables people to understand themselves. Behavioral sciences — exploring areas such as human drives and motivation, the differences between humans and other animals, and group dynamics — now cover a wide body of literature. Unfortunately, managers often have surprisingly little interest in such research. However, education essential to the development of mature individuals *begins* with a reevaluation of the self and what it means to be human.

Management's second responsibility is to provide a favorable work environment by eliminating the psychological and physical environmental problems that negatively affect workers — problems that may significantly hamper training designed to develop worker autonomy.

To create a favorable psychological environment requires first escape from an authoritarian management system and then changes in company structure to promote participative management. William G. Ouichi (*Theory Z*) compared Japanese and American management styles and discovered that management techniques similar to those of the Japanese had been successfully implemented in many leading U.S. companies. He labelled these companies type-Z firms. Their most distinctive characteristic was a management commitment to employees, which provided the foundation for mutual trust, concern, and egalitarianism. Ouichi concluded that the Japanese do not have a monopoly on such practices.

Likert and Ouichi both have argued that respect for workers and a company structure that supports employees help develop autonomous workers and create a psychological environment that encourages small group activities.

The work environment is further enhanced when management establishes certain physical conditions, such as an autonomous maintenance organizational structure or an improved factory environment. For example, if workers are encouraged to carry out small group activities but have no suitable place to hold meetings, they will most likely lose enthusiasm. Some groups, however, lacking an appropriate meeting place, have held their meetings outdoors. Companies where top management is enthusiastic about small group activities often build a lounge in the factory for use as a meeting room.

TPM cannot be implemented if top management fails to provide the psychological and physical environment that promotes true participative management.

Factories that have won the PM Prize have proved that TPM is profitable. Every company has an untapped source of potential profits that can be activated through TPM. Moreover, a company that achieves prize-winning levels in TPM implementation has gained the skilled work force, the equipment, and the factory environment necessary to survive in the age of automation and unmanned operations.

Appendix
The PM Prize for
Outstanding TPM Plants

In Japan, the Distinguished Plant Prize (or PM Prize) is awarded annually to plants that successfully implement TPM. Due to increased public attention, this award is now as highly coveted as the Deming Prize, which has been awarded for exemplary quality programs and achievements for the past 30 years. The PM prize is awarded by the Japan Institute of Plant Maintenance (JIPM), a key promoter of TPM. A special PM Prize committee within JIPM selects the prize winners.

The PM Prize has been awarded since 1964. While early prize winners were selected for their outstanding productive maintenance, implementation of TPM has become a requirement for the PM Prize since 1971, when Nippondenso Co., Ltd. became the first firm in Japan to win the PM Prize for TPM.

The PM Prize is offered in two categories, for large corporations and firms with fewer than 1,000 employees and less than ¥500 million ($2.2 million) in capital assets. The PM Prize Committee examines the concrete results achieved by applicants through TPM implementation. Therefore, progress in areas such as systematization or standardization is disregarded. Selection of winners is based on improvements achieved through proper equipment maintenance, *i.e.*, increased productivity and quality, reduced costs, reduced inventory, elimination of accidents, pollution control, and the creation of a favorable work environment. Each year, standards (based on actual results) are going up.

Table A lists the PM Prize winners since 1971, when TPM implementation became a requirement for consideration. By 1982, 29 companies in category 1 and 22 companies in category 2 had been awarded the PM Prize for their TPM. The dramatic increase in prize-winners after 1981 is noteworthy.

APPLICATION PROCEDURES FOR THE PM PRIZE

To be considered for the PM Prize, a company must compile a "TPM Implementation Report" and submit it to the PM Prize Committee. Although compiling this report is a formidable task, it forces companies to reevaluate their stages of TPM implementation. This may lead to the discovery of hidden weaknesses and result in additional improvements, and many of the prize winners described it as a perfect opportunity for a comprehensive survey of their equipment maintenance program.

A company being considered for the PM Prize will be evaluated by specialists in the related fields who will point out weaknesses in their programs and suggest improvements. Applicants invariably find this helpful, because the advice received helps boost the future level of TPM.

The PM Prize is awarded once a year. Judging occurs in two stages, the initial screening (based on documents) and a second screening (based on a factory visit). The "TPM Implementation Report" must be submitted to the JIPM by the end of May of a given year for the first screening, which takes place at a committee meeting in June. When a company has passed this screening, the PM Prize Committee will notify it of the exact date for the factory examination, usually some time between mid-July and mid-August. All results are evaluated by early September. Then, the PM Prize Committee consisting of the JIPM chairperson and other experts selects the winners. The award ceremony is held in late September or early October at the National Conference on Equipment Maintenance.

For Japanese companies, the PM Prize symbolizes a new beginning and a challenge to strive for even greater improvement. Frequently, representatives from PM prize-winning workplaces are invited to report on their continuing efforts at lectures and conferences, and JIPM sponsors field trips to their plants. Both of these practices are incentives for continued improvement. Repeated presentation of company results to the public and frequent visitors to the plant help employees strive for higher and more challenging goals.

The award level beyond the PM Prize is the Special (Distinguished) PM Prize. Winners of this prize have eliminated the weaknesses discovered earlier by the PM Prize Committee and developed unique maintenance techniques and equipment technology; in other words, they have succeeded in raising their standards of TPM.

Year Category	Category 1	Category 2
1971	Nippondenso Mitsubishi Heavy Industries	(none)
1972	Toray Industries (Okazaki plant)	Sumiko I.S.P. Co. (Hama plant)
1973	Toyoda Gōsei	Chūsei Rubber Segawa Chemical Industries Hinode Rubber Industries Chūo Rubber Industries Suzuki Chemical Industries
1974	(none)	Ichiei Industries Hokusei Rubber Kitano Manufacturing
1975	Fuji Photo Film (Odawara plant)	Toyokuni Industries
1976	Special Prize: Toyoda Gōsei Kawasaki Steel (Chiba plant) Yokohama Rubber (Mishima plant)	Special Prize: Suzuki Chemical Industries Kaiyō Rubber Shiota Kasei Co. Yokoyama Spring
1977	Sumitomo Metal Industries (Wakayama plant) Fuji Photo Film (Fujinomiya plant) Yokohama Rubber (Mie plant) Wakō Chemical Industries	Yamako Co.

Note: Category 2 consists of companies with less than ¥500 million ($2 million) in capital assets and less than 1,000 employees.

Year Category	Category 1	Category 2
1978	Chūō Spring (Hekinan plant)	Anjo Denki
1979	Aisan Industries Toyota Steel Works	(none)
1980	Aichi Steel Works	(none)
1981	Anjo Denki Topy Industries (Kanagawa Plant) Tōkai Rubber Industries (Komaki plant) Tokyū Car Corporation (Osaka plant) Maruyasu Kōgyō (Okazaki plant) Matsushita Electric Industries Mikuni plant)	Kyōwa Precision Teikei Machine Matsuo Seisakusho Miyama Steel
1982	Aishin Seiki Co. Central Motor Wheel (Toyada plant) Nippon Zeon Co. (Kawasaki, Takaoka, Tokuyama, and Mizushima plants) Fuji Photo Film (Yoshida Minami plant)	Tōhoku Satake Seisakusho

Note: Category 2 consists of companies with less than ¥ 500 million ($4 million) in capital assets and less than 1,000 employees.

Table A. PM Prize-Winning Companies (from 1971)

About the Author

Seiichi Nakajima graduated with a degree in mechanical engineering from Kanazawa Technical College in 1939. He joined the Japan Management Association in 1949 and since then has served as a management consultant for more than one hundred companies. Mr. Nakajima introduced PM to Japan in 1951 and has remained its leading advocate and educator over the past thirty years. He is currently Vice Chairman of the Japan Institute of Plant Maintenance as well as Executive Vice President of Japan Management Association.

Mr. Nakajima has written many books and articles on the subject of plant engineering and management. *Introduction to TPM* is the first to be published in English.

Index

BOOKS FROM PRODUCTIVITY PRESS

Productivity Press publishes books that empower individuals and companies to achieve excellence in quality, productivity, and the creative involvement of all employees. Through steadfast efforts to support the vision and strategy of continuous improvement, Productivity Press delivers today's leading-edge tools and techniques gathered directly from industrial leaders around the world. Call toll-free 1-800-394-6868 for our free catalog.

5 Pillars of the Visual Workplace
The Sourcebook for 5S Implementation
Hiroyuki Hirano

In this important sourcebook recently published by Productivity Press, JIT expert Hiroyuki Hirano provides the most vital information available on the visual workplace. He describes the 5S's: seiri, seiton, seiso, seiketsu, shitsuke (which translate as organization, orderliness, cleanliness, standardized cleanup, and discipline). Hirano discusses how the 5S theory fosters efficiency, maintenance, and continuous improvement in all areas of the company, from the plant floor to the sales office. Presented in a thorough, detailed style, *5 Pillars of the Visual Workplace* explains why the 5S's are important and the who, what, where, and how of 5S implementation. This book includes numerous case studies, hundreds of graphic illustrations, and over forty 5S user forms and training materials.
ISBN 1-56327-047-1 / 353 pages, illustrated / $85.00 / Order FIVE-B156

20 Keys to Workplace Improvement
Iwao Kobayashi

The 20 Keys system does more than just bring together twenty of the world's top manufacturing improvement approaches—it integrates these individual methods into a closely interrelated system for revolutionizing every aspect of your manufacturing organization. This revised edition of Kobayashi's bestseller amplifies the synergistic power of raising the levels of all these critical areas simultaneously. The new edition presents upgraded criteria for the five-level scoring system in most of the 20 Keys, supporting your progress toward becoming not only best in your industry but best in the world. New material and an updated layout throughout assist managers in implementing this comprehensive approach. In addition, valuable case studies describe how Morioka Seiko (Japan) advanced in Key 18 (use of microprocessors) and how Windfall Products (Pennsylvania) adapted the 20 Keys to its situation with good results.
ISBN 1-56327-109-5 / 312 pages / $50.00 / Order 20KREV-B156

PRODUCTIVITY PRESS, INC., DEPT. BK, P.O. BOX 13390, PORTLAND, OR 97213-0390
Telephone: 1-800-394-6868   Fax: 1-800-394-6286

Becoming Lean
Inside Stories of U.S. Manufacturers
Jeffrey Liker

Most other books on lean management focus on technical methods and offer a picture of what a lean system should look like. Some provide snapshots of before and after. This is the first book to provide technical descriptions of successful solutions and performance improvements. The first book to include powerful first-hand accounts of the complete process of change, its impact on the entire organization, and the rewards and benefits of becoming lean. At the heart of this book you will find the stories of American manufacturers who have successfully implemented lean methods. Authors offer personalized accounts of their organization's lean transformation, including struggles and successes, frustrations and surprises. Now you have a unique opportunity to go inside their implementation process to see what worked, what didn't, and why. Many of these executives and managers who led the charge to becoming lean in their organizations tell their stories here for the first time!

ISBN 1-56327-173-7/ 350 pages / $35.00 / Order LEAN-B156

CEDAC
A Tool for Continuous Systematic Improvement
Ryuji Fukuda

CEDAC, encompasses three tools for continuous systematic improvement: window analysis (for identifying problems), the CEDAC diagram (a modification of the classic "fishbone diagram," for analyzing problems and developing standards), and window development (for ensuring adherence to standards). This manual provides directions for setting up and using CEDAC. Sample forms included.

ISBN 1-56327-140-0 / 144 pages / $30.00 / Order CEDAC-B156

Corporate Diagnosis
Setting the Global Standard for Excellence
Thomas L. Jackson with Constance E. Dyer

All too often, strategic planning neglects an essential first step and final step-diagnosis of the organization's current state. What's required is a systematic review of the critical factors in organizational learning and growth, factors that require monitoring, measurement, and management to ensure that your company competes successfully. This executive workbook provides a step-by-step method for diagnosing an organization's strategic health and measuring its overall competitiveness against world class standards. With checklists, charts, and detailed explanations, Corporate Diagnosis is a practical instruction manual. The pillars of Jackson's diagnostic system are strategy, structure, and capability. Detailed diagnostic questions in each area are provided as guidelines for developing your own self-assessment survey.

ISBN 1-56327-086-2 / 115 pages / $65.00 / Order CDIAG-B156

PRODUCTIVITY PRESS, INC., DEPT. BK, P.O. BOX 13390, PORTLAND, OR 97213-0390
Telephone: 1-800-394-6868   Fax: 1-800-394-6286

Cycle Time Management
The Fast Track to Time-Based Productivity Improvement
Patrick Northey and Nigel Southway

As much as 90 percent of the operational activities in a traditional plant are nonessential or pure waste. This book presents a proven methodology for eliminating this waste within 24 to 30 months by measuring productivity in terms of time instead of revenue or people. CTM is a cohesive management strategy that integrates just-in-time (JIT) production, computer integrated manufacturing (CIM), and total quality control (TQC). From this succinct, highly focused book, you'll learn what CTM is, how to implement it, and how to manage it.
ISBN 1-56327-015-3 / 200 pages / $35.00 / Order CYCLE-B156

Implementing a Lean Management System
Thomas L. Jackson with Karen R. Jones

Does your company think and act ahead of technological change, ahead of the customer, and ahead of the competition? Thinking strategically requires a company to face these questions with a clear future image of itself. Implementing a Lean Management System lays out a comprehensive management system for aligning the firm's vision of the future with market realities. Based on hoshin management, the Japanese strategic planning method used by top managers for driving TQM throughout an organization, Lean Management is about deploying vision, strategy, and policy to all levels of daily activity. It is an eminently practical methodology emerging out of the implementation of continuous improvement methods and employee involvement. The key tools of this book build on multiskilling, the knowledge of the worker, and an understanding of the role of the new lean manufacturer.
ISBN 1-56327-085-4 / 182 pages / $65.00 / Order ILMS-B156

Implementing TPM
The North American Experience
Charles J. Robinson and Andrew P. Ginder

The authors document an approach to TPM planning and deployment that modifies the JIPM 12-step proc-ess to accommodate the experiences of North American plants. They include details and advice on specific deployment steps, OEE calculation methodology, and autonomous maintenance deployment. This book shows how to make TPM work in unionized plants and how to position TPM to support and complement other strategic manufacturing improvement initiatives.
ISBN 1-56327-087-0 / 224 pages / $45.00 / Order IMPTPM-B156

PRODUCTIVITY PRESS, INC., DEPT. BK, P.O. BOX 13390, PORTLAND, OR 97213-0390
Telephone: 1-800-394-6868   Fax: 1-800-394-6286

Integrating Kanban with MRPII
Automating a Pull System for Enhanced JIT Inventory Management
Raymond S. Louis

Manufacturing organizations continuously strive to match the supply of products to market demand. Now for the first time, the automated kanban system is introduced utilizing MRPII. This book describes an automated kanban system that integrates MRPII, kanban bar codes and a simple version of electronic data interchange into a breakthrough system that substantially lowers inventory and significantly eliminates non-value adding activities. This new system automatically recalculates and triggers replenishment, integrates suppliers into the manufacturing loop, and uses bar codes to enhance speed and accuracy of the receipt process. From this book, you will learn how to enhance the flexibility of your manufacturing organization and dramatically improve your competitive position.

ISBN 1-56327-182-6 / 200 pages / $45.00 / Item # INTKAN-B156

JIT Factory Revolution
A Pictorial Guide to Factory Design of the Future
Hiroyuki Hirano

The first encyclopedic picture-book of Just-In-Time, using photos and diagrams to show exactly how JIT looks and functions in production and assembly plants. Unprecedented behind-the-scenes look at multiprocess handling, cell technology, quick changeovers, kanban, andon, and other visual control systems. See why a picture is worth a thousand words.

ISBN 0-915299-44-5 / 218 pages / $50.00 / Order JITFAC-B156

Kaizen for Quick Changeover
Going Beyond SMED
Kenichi Sekine and Keisuke Arai

Especially useful for manufacturing managers and engineers, this book describes exactly how to achieve faster changeover. Picking up where Shingo's SMED book left off, you'll learn how to streamline the process even further to reduce changeover time and optimize staffing at the same time.

ISBN 0-915299-38-0 / 315 pages / $75.00 / Order KAIZEN-B156

PRODUCTIVITY PRESS, INC., DEPT. BK, P.O. BOX 13390, PORTLAND, OR 97213-0390
Telephone: 1-800-394-6868   Fax: 1-800-394-6286

Manufacturing Strategy
How to Formulate and Implement a Winning Plan
John Miltenburg

This book offers a step-by-step method for creating a strategic manufacturing plan. The key tool is a multidimensional worksheet that links the competitive analysis to manufacturing outputs, the seven basic production systems, the levels of capability and the levers for moving to a higher level. The author presents each element of the worksheet and shows you how to link them to create an integrated strategy and implementation plan. By identifying the appropriate production system for your business, you can determine what output you can expect from manufacturing, how to improve outputs, and how to change to more optimal production systems as your business needs changes. This is a valuable book for general managers, operations managers, engineering managers, marketing managers, comptrollers, consultants, and corporate staff in any manufacturing company.
ISBN 1-56327-071-4 / 391 pages / $45.00 / Order MANST-B156

Modern Approaches to Manufacturing Improvement
The Shingo System
Alan Robinson (ed.)

Here's the quickest and most inexpensive way to learn about the pioneering work of Shigeo Shingo, co-creator (with Taiichi Ohno) of Just-In-Time. It's an introductory book containing excerpts of five of his classic books as well as an excellent introduction by Professor Robinson. Learn about quick changeover, mistake-proofing (poka-yoke), non-stock production, and how to apply Shingo's "scientific thinking mechanism."
ISBN 0-915299-64-X / 420 pages / $23.00 paper / Order READER-B156

One-Piece Flow
Cell Design for Transforming the Production Process
Kenichi Sekine

By reconfiguring your traditional assembly lines into production cells based on one-piece flow, you can drastically reduce your lead time, manpower requirements, and number of defects. Sekine examines the basic principles of process flow building, then offers detailed case studies of how various industries designed unique one-piece flow systems to meet their particular needs.
ISBN 0-915299-33-X / 308 pages / $75.00 / Order 1PIECE-B156

PRODUCTIVITY PRESS, INC., DEPT. BK, P.O. BOX 13390, PORTLAND, OR 97213-0390
Telephone: 1-800-394-6868   Fax: 1-800-394-6286

Poka-Yoke
Improving Product Quality by Preventing Defects
Nikkan Kogyo Shimbun Ltd. and Factory Magazine (ed.)

If your goal is 100 percent zero defects, here is the book for you—a completely illustrated guide to poka-yoke (mistake-proofing) for supervisors and shop-floor workers. Many poka-yoke devices come from line workers and are implemented with the help of engineering staff. The result is better product quality—and greater participation by workers in efforts to improve your processes, your products, and your company as a whole.
ISBN 0-915299-31-3 / 295 pages / $65.00 / Order IPOKA-B156

A Revolution in Manufacturing
The SMED System
Shigeo Shingo

The heart of JIT is quick changeover methods. Dr. Shingo, inventor of the Single-Minute Exchange of Die (SMED) system for Toyota, shows you how to reduce your changeovers by an average of 98 percent! By applying Shingo's techniques, you'll see rapid improvements (lead time reduced from weeks to days, lower inventory and warehousing costs) that will improve quality, productivity, and profits.
ISBN 0-915299-03-8 / 383 pages / $75.00 / Order SMED-B156

A Study of the Toyota Production System from an Industrial Engineering Viewpoint
Shigeo Shingo

Here is Dr. Shingo's classic industrial engineering rationale for the priority of process-based over operational improvements for manufacturing. He explains the basic mechanisms of the Toyota production system in a practical and simple way so that you can apply them in your own plant. This book clarifies the fundamental principles of JIT including leveling, standard work procedures, multi-machine handling, and more.
ISBN 0-915299-17-8 / 291 pages / $50.00 / Order STREV-B156

TPM in Process Industries
Tokutaro Suzuki (ed.)

Process industries have a particularly urgent need for collaborative equipment management systems like TPM that can absolutely guarantee safe, stable operation. In *TPM in Process Industries*, top consultants from JIPM (Japan Institute of Plant Maintenance) document approaches to implementing TPM in process industries. They focus on the process environment and equipment issues such as process loss structure and calculation, autonomous maintenance, equipment and process improvement, and quality maintenance. Must reading for any manager in the process industry.
ISBN 1-56327-036-6 / 400 pages / $85.00 / Order TPMPI-B156

PRODUCTIVITY PRESS, INC., DEPT. BK, P.O. BOX 13390, PORTLAND, OR 97213-0390
Telephone: 1-800-394-6868   Fax: 1-800-394-6286

Visual Feedback Photography
Making Your 5S Implementation Click
Adapted from materials by Ken'ichi Ono

Are you looking for a way to breath some life into your 5S activities in a way that vividly demonstrates your progress? Consider capturing the evolution of your program in photographs. Visual Feedback Photography is a simple method for teams to use as they implement workplace improvements, and a means to record changes in the workplace over time. The result is a series of photographs displayed on a workplace chart, providing a clear record of improvement activities related to workplace problem areas.
ISBN 1-56327-090-1 /$150.00 / Order VFPACT-B156

Zero Quality Control
Source Inspection and the Poka-Yoke System
Shigeo Shingo

Dr. Shingo reveals his unique defect prevention system, which combines source inspection and poka-yoke (mistake-proofing) devices that provide instant feedback on errors before they can become defects. The result: 100 percent inspection that eliminates the need for SQC and produces defect-free products without fail. Includes 112 examples, most costing under $100. Two-part video program also available; call for details.
ISBN 0-915299-07-0 / 328 pages / $75.00 / Order ZQC-B156

PRODUCTIVITY PRESS, INC., DEPT. BK, P.O. BOX 13390, PORTLAND, OR 97213-0390
Telephone: 1-800-394-6868   Fax: 1-800-394-6286

TO ORDER: Write, phone, or fax Productivity Press, Dept. BK, P.O. Box 13390, Portland, OR 97213-0390, phone 1-800-394-6868, fax 1-800-394-6286. Send check or charge to your credit card (American Express, Visa, MasterCard accepted).

U.S. ORDERS: Add $5 shipping for first book, $2 each additional for UPS surface delivery. Add $5 for each AV program containing 1 or 2 tapes; add $12 for each AV program containing 3 or more tapes. We offer attractive quantity discounts for bulk purchases of individual titles; call for more information.

ORDER BY E-MAIL: Order 24 hours a day from anywhere in the world. Use either address:
    To order: service@ppress.com
    To view the online catalog and/or order: http://www.ppress.com/

QUANTITY DISCOUNTS: For information on quantity discounts, please contact our sales department.

INTERNATIONAL ORDERS: Write, phone, or fax for quote and indicate shipping method desired. For international callers, telephone number is 503-235-0600 and fax number is 503-235-0909. Prepayment in U.S. dollars must accompany your order (checks must be drawn on U.S. banks). When quote is returned with payment, your order will be shipped promptly by the method requested.

NOTE: Prices are in U.S. dollars and are subject to change without notice.

About the Shopfloor Series

Put powerful and proven improvement tools in the hands of your entire workforce!

Progressive shopfloor improvement techniques are imperative for manufacturers who want to stay competitive and to achieve world class excellence. And it's the comprehensive education of all shopfloor workers that ensures full participation and success when implementing new programs. The Shopfloor Series books make practical information accessible to everyone by presenting major concepts and tools in simple, clear language and at a reading level that has been adjusted for operators by skilled instructional designers. One main idea is presented every two to four pages so that the book can be picked up and put down easily. Each chapter begins with an overview and ends with a summary section. Helpful illustrations are used throughout.

Books currently in the Shopfloor Series include:

5S FOR OPERATORS
5 Pillars of the Visual Workplace
The Productivity Press Development Team
ISBN 1-56327-123-0
incl. application questions / 133 pages
Item # 5SOP-B156 / $25.00

QUICK CHANGEOVER FOR OPERATORS
The SMED System
The Productivity Press Development Team
ISBN 1-56327-125-7
incl. application questions / 93 pages
Item # QCOOP-B156/ $25.00

MISTAKE-PROOFING FOR OPERATORS
The Productivity Press Development Team
ISBN 1-56327-127-3 / 93 pages
Item # ZQCOP-B156/ $25.00

TPM FOR SUPERVISORS
The Productivity Press Development Team
ISBN 1-56327-161-3 / 96 pages
Item # TPMSUP-B156/ $25.00

TPM TEAM GUIDE
Kunio Shirose
ISBN 1-56327-079-X / 175 pages
Item # TGUIDE-B156 / $25.00

TPM FOR EVERY OPERATOR
Japan Institute of Plant Maintenance
ISBN 1-56327-080-3 / 136 pages
Item # TPMEO-B156 / $25.00

AUTONOMOUS MAINTENANCE
Japan Institute of Plant Maintenance
ISBN 1-56327-081-1 / 138 pages
Item # AUTMOP-B156 / $25.00

FOCUSED EQUIPMENT IMPROVEMENT
Japan Institute of Plant Maintenance
ISBN 1-56327-081-1 / 138 pages
Item # FEIOP-B156 / $25.00

Productivity Press, Dept. BK, P.O. Box 13390, Portland, OR 97213-0390
Telephone: 1-800-394-6868   Fax: 1-800-394-6286

Continue Your Learning with In-House Training and Consulting from the Productivity Consulting Group

The Productivity Consulting Group (PCG) offers a diverse menu of consulting services and training products based on the exciting ideas contained in the books of Productivity Press. Whether you need assistance with long term planning or focused, results-driven training, PCG's experienced professional staff can enhance your pursuit of competitive advantage.

PCG integrates a cutting edge management system with today's leading process improvement tools for rapid, measurable, lasting results. In concert with your management team, PCG will focus on implementing the principles of Value Adding Management, Total Quality Management, Just-In-Time, and Total Productive Maintenance. Each approach is supported by Productivity's wide array of team-based tools: Standardization, One-Piece Flow, Hoshin Planning, Quick Changeover, Mistake-Proofing, Kanban, Problem Solving with CEDAC, Visual Workplace, Visual Office, Autonomous Maintenance, Equipment Effectiveness, Design of Experiments, Quality Function Deployment, Ergonomics, and more. And, based on the continuing research of Productivity Press, PCG expands its offering every year.

Productivity is known for significant improvement on the shopfloor and the bottom line. Through years of repeat business, an expanding and loyal client base continues to recommend Productivity to their colleagues. Contact PCG to learn how we can tailor our services to fit your needs.

Telephone: 1-800-966-5423 (U.S. only) or 1-203-846-3777
Fax: 1-203-846-6883